Value Engineering in the Construction Industry

ALPHONSE J. DELL'ISOLA

A Construction Publishing Company Book

VNR **VAN NOSTRAND REINHOLD COMPANY**
NEW YORK   CINCINNATI   ATLANTA   DALLAS   SAN FRANCISCO
LONDON   TORONTO   MELBOURNE

Van Nostrand Reinhold Company Regional Offices:
New York   Cincinnati   Atlanta   Dallas   San Francisco

Van Nostrand Reinhold Company International Offices:
London   Toronto   Melbourne

Library of Congress Catalog Card Number: 77-4930
ISBN: 0-442-12153-9, formerly carried as ISBN 0-913634-10-7

Manufactured in the United States of America

Published by Van Nostrand Reinhold Company
450 West 33rd Street, New York, N.Y. 10001

Published simultaneously in Canada by Van Nostrand Reinhold Ltd.

15 14 13 12 11 10 9 8 7 6 5 4 3 2 1

Library of Congress Cataloging in Publication Data

Dell'Isola, Alphonse J
    Value engineering in the construction industry.

    "A Construction Publishing Company book."
    Bibliography: p.
    Includes index.
    1. Construction industry—Costs.   2. Value
analysis (cost control)   I. Title.
TH437.D43   1975   658.1′552   77-4930
ISBN 0-442-12153-9

PREFACE

This guide outlines the methodology developed by the author as a result of applying value analysis/engineering (VA/E) principles to construction projects. The information contained in the guide is based on experience dating back to 1962 when the author pioneered the first formal application of VA/E to construction.

Acknowledgments are made to the many contributors of information and moral support in the Department of Defense, the Naval Facilities Engineering Command, the first construction agency to establish a VA/E program, and Louis C. Kingscott, AIA, of Kalamazoo, Michigan.

Special acknowledgment is due my secretary, Nancy Crist, who labored through various drafts, Jules Godel and Dorothy Obre who helped with editing, and my wife, who spent many lonely nights while the manuscript was being prepared.

The guide is a positive approach to effecting economies in construction in an era of rising costs and budgetary pressures.

CONTENTS

INTRODUCTION

The construction industry is the largest industry in the United States, with approximately $80 billion in annual expenditures. Its history has been one of extreme competitiveness, with margins of profit generally low compared with the risks involved and with other areas of the economy. The cost of building products is continually rising and wages are consistently higher than those of other industries.

As a control over the industry, building codes are used. These are national and local codes, which in most cases were established by trial and error and are compilations of previously proved practices. In many areas these codes have not kept pace with new technological advances. For example, plastic pipe is used industrially to a great extent; yet, many codes still prohibit its use.

Let's take a deeper look into the present methods used by the construction industry. Generally, a facility is designed by an architect and engineer under a negotiated contract. Construction is performed by a contractor selected through competitively bid, fixed-price contracts. However, large industrial firms, such as DuPont and General Motors, conduct their own design in-house and then issue fixed-price contracts to a contractor selected through competitive bidding from a selected list of bidders. In addition, there are a number of "turnkey" contractors who design, engineer, and construct a complete facility according to the owner's need at a fixed price.

Government methods vary somewhat from those of the private sector. Basically, the design of a facility is negotiated with an architect-engineering firm, and the construction is performed by a contractor selected by competitive bidding. A basic difference is that government agencies are required to open bidding to all contractors, while the private sector generally uses a select bidder's list. Some government agencies, such as the Bureau of Reclamation and the Bureau of Indian Affairs, do most of their own design, but have construction performed by others. Construction inspection is normally done by the government agency; however, there are times when this responsibility is added to the architect and engineer contract. The Post Office Department has had considerable work performed under the "lease" plan. Under this plan, the department specifies requirements and solicits a long-term lease from a private owner for a suitable facility. The owner builds the facility to meet the department's requirements.

The advent of the systems concept, the rising cost of money, and the increasing importance of time have forced a reappraisal of present methods. More turnkey projects are being tried, and a short-cut method of design and build, sometimes called "fast track" or project delivery system/scheduling, is becoming more prevalent. For a graphic portrayal of the shortened method, see Figure 1. This method permits the owner to pre-bid portions of the work, and the contractor to order material and start portions of the work concurrently with development of the design drawings, thus gaining valuable time over the conventional process. This method requires careful selection of both the designer and contractor, as close cooperation between them is essential.

Some government agencies (the General Services Administration and the Department of Health, Education, and Welfare) use a construction manager, who is responsible for a project from design to occupancy. He is given authority to issue subcontracts for various areas of work and to monitor all construction.

These new trends are reducing the 3- to 5-year period between concept and occupancy that is required by existing construction methods. The result is overall savings to the owner. See Figure 2 for a time comparison between the federal government and private industry for designing and constructing a large office building.* The significant time difference is caused by the myriad approvals required, approval agents involved, government laws enforced, and general insensitivity to time requirements in federal government procedures.

Figure 3 graphically portrays the trend in construction costs over the last 60 years. An interesting comparison can be noted. Construction prices have almost doubled in 10 years, while the cost of manufactured goods such as cars, television sets, etc., has risen comparatively little. The principal reason is that in the industrialized sector of our economy, rising prices are offset by increased productivity using new methods and materials. Who in the construction industry is really concerned about rising costs?

The architect-engineer's fees are based on a percentage of construction costs. The more a facility costs, the more he makes. The contractor's profit is also based on a percentage of construction costs. Again, the more it costs, the better chance he has of making more money. It is the owner who bears the brunt of the inflationary

*Construction Contracting Systems; A Report on the Systems Used by PBS and Other Organizations, is available from Public Buildings Service, General Services Administration, Washington, D.C. 20405.

spiral, and he has done little to bring about controls over the various factions involved, especially field labor. What is needed is an owner's revolt to force the construction cost index to be more in line with the industrial segment of our economy.

It is believed that use of the methodology outlined in this book offers one real tool to help the construction industry reduce costs.

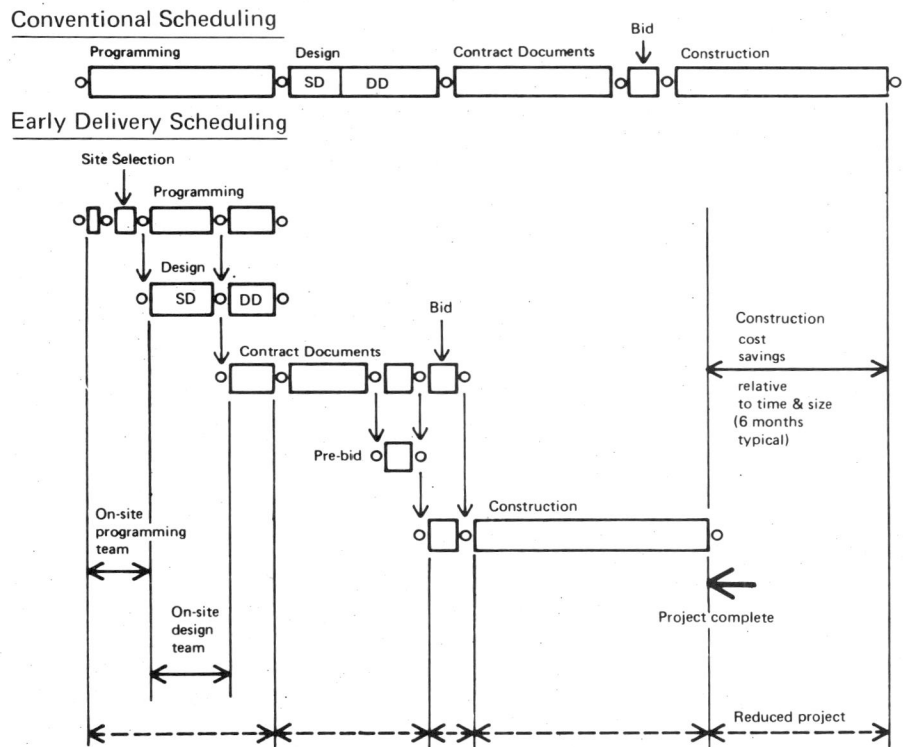

FIGURE 1.
Project delivery system/scheduling. SD, schematic design; DD, design development; programming, developing scope of project; prebid, long-lead items component systems; bid, construction contract solicitation.

Preliminary Planning

Site and architect selection

Invitation for Bids

Design (16 mo.) and design review (8 mo.)

Construction

Variable | 5 | 24 | 5 | 30
59
Months

PRIVATE INDUSTRY

Invitation for Bids

Preliminary planning — site & architect selection

Design and design review

Construction

Time difference

Variable | 10 | 2 | 12 | 35
24
Months

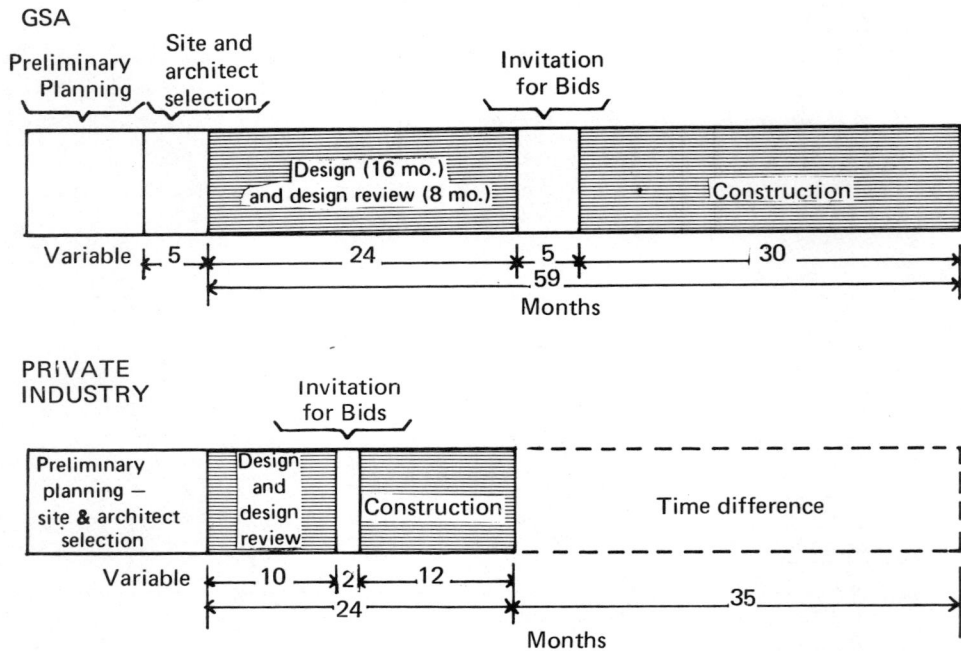

FIGURE 2. **Average time for GSA and private industry to design and build a $10 million office building.**

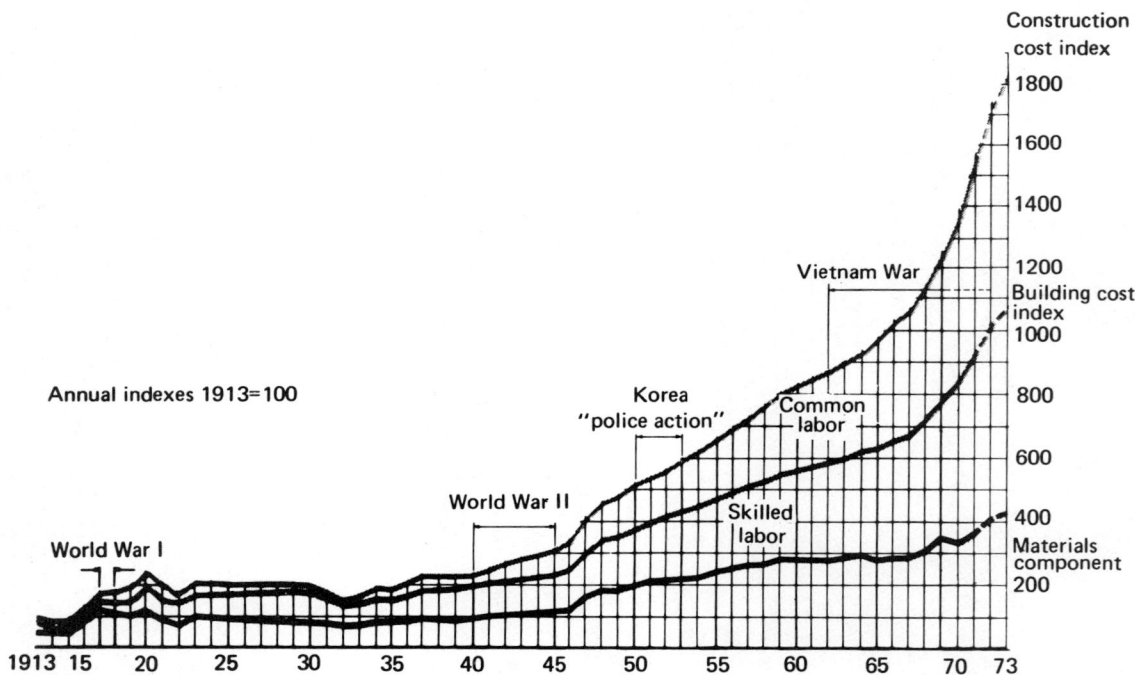

Construction cost index

1800
1600
1400
1200
Building cost index
1000
800
600
400
Materials component
200

Annual indexes 1913=100

Vietnam War

Korea "police action"

Common labor

World War II

Skilled labor

World War I

1913 15   20   25   30   35   40   45   50   55   60   65   70 73

FIGURE 3. *Engineering News-Record* **indexes of basic construction costs.**

THE VALUE CONCEPT

History

Value analysis evolved during World War II when shortages of materials and labor forced the introduction of many substitutes. The management of the General Electric Company noted that often these substitutes reduced costs and improved the product. So they asked Harry Erlicher the question—Why? He wondered if something that had happened by accident could be made to happen on purpose. Lawrence Miles, who worked for Mr. Erlicher, was called upon to answer the question. Miles developed a system of techniques, which he called value analysis, that made significant improvements in a product systematically rather than by accident.

As these methods were adapted to other productive processes, the name was changed to value engineering. Today the two names—value engineering (VE) and value analysis (VA)—are used synonymously. Terms such as value control, and value management are also used.

The Department of Defense adopted VE in 1954 when the Navy's Bureau of Ships applied the VE concept to its procurement actions. Secretary Robert S. McNamara, in 1964, expanded the cost-reduction program which led to further utilization of VE principles. At that time, many federal, state, and local government agencies adopted or considered the advantages of formal VE programs in their activities. Additionally, many industrial companies established formal VE programs as part of their profit improvement efforts. As its use became more widespread, a formal definition was required.

Definition

In simple terms, VE is a systematic approach to obtaining optimum value for every dollar spent. Through a system of investigation, unnecessary expenditures are avoided, resulting in improved value and economy. The VE approach is a creative effort directed toward the analysis of functions. It is concerned with the elimination or modification of anything that adds cost to an item without adding to its function. During this process all expenditures relating to construction, maintenance, operation, replacement, etc., are considered.

Through the use of creative techniques and the latest technical information regarding new materials and methods, alternate solutions are developed for the specific function. In contrast to costcutting by simply making smaller quantities or using fewer or cheaper materials, VE analyzes function or method by asking such questions as:

> What is it?
> What does it do?
> What must it do?
> What does it cost?
> What other material or method could be used to do the same job?
> What would the alternate material or method cost?

The following definition and quotations are taken from Miles' book.*

Value analysis/engineering is an organized, creative approach which has for its purpose the efficient identification of unnecessary costs, i.e., costs which provide neither quality nor use nor life nor appearance nor customer features.

Value analysis is *not* a substitute for conventional cost reduction work methods. Rather, it is a potent and completely different procedure for accomplishing far greater results. It improves the effectiveness of work that has been conventionally performed over the years, as it fills in blind spots. Quite commonly 15 to 25 percent, and very often much more, of the manufacturing costs can be removed by effective application of the teaching of value analysis.

Too often in the past, an endeavor to remove cost without the use of professional tools for accomplishing the project has resulted in a lowering of quality. Therefore, it must be clearly understood from the start that accomplishing better value does not mean reducing quality to a point where it is lower but may just get by. No reduction whatever in needed quality is tolerated in the professional grade of value work.

Experience shows that quality is frequently increased as the result of developing alternatives for accomplishment of the use of esteem functions.

Total Cost of Facilities

VE in construction is an objective, systematic method of optimizing the total cost of a facility or system for a specific number of years. Total cost means the ultimate costs to construct, operate, maintain, and replace a facility or system over a specific life cycle.

*L. D. Miles. *Techniques of Value Analysis and Engineering,* 2nd ed. New York, McGraw-Hill, 1961.

TOTAL COST CONCEPT

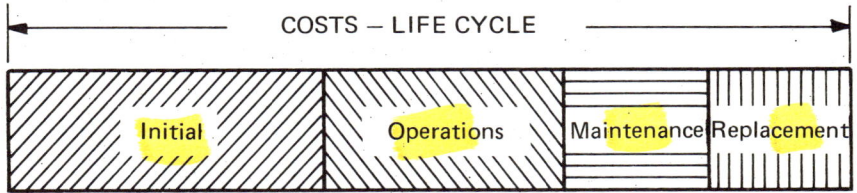

TOTAL COST DISTRIBUTION

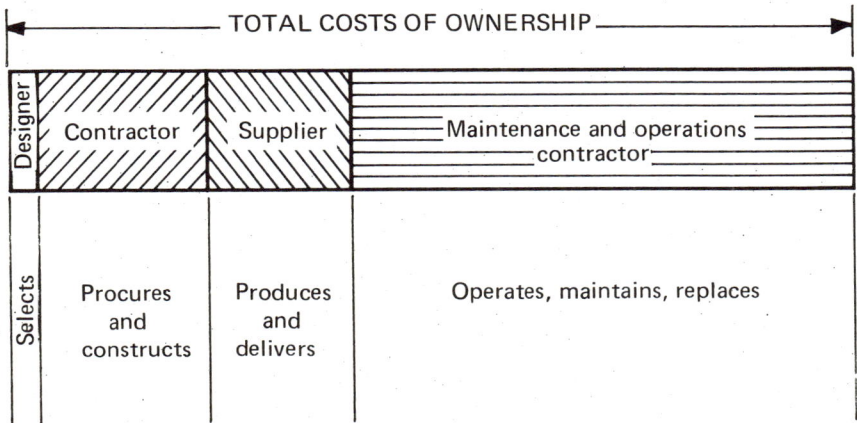

FIGURE 4. Total cost concept and total cost distribution in facilities construction.

Figure 4 illustrates both the relative breakdown of all costs for a typical multi-million-dollar hospital project over its normal life cycle and the distribution of the using agency's money over the same period. It is interesting to note that the designer's portion represents the smallest monetary area. This fact warrants some thought, especially since his decisions have the greatest impact on total costs. The cost of the facility's system operational personnel (doctors, nurses, technicians, hospital administrators, etc.) is not included. It is also interesting to note that the breakdown shown in Figure 4 would closely approximate ownership costs for a home or motor vehicle. Normally initial costs (design, contractor, supplier) equal approximately one-third of total costs.

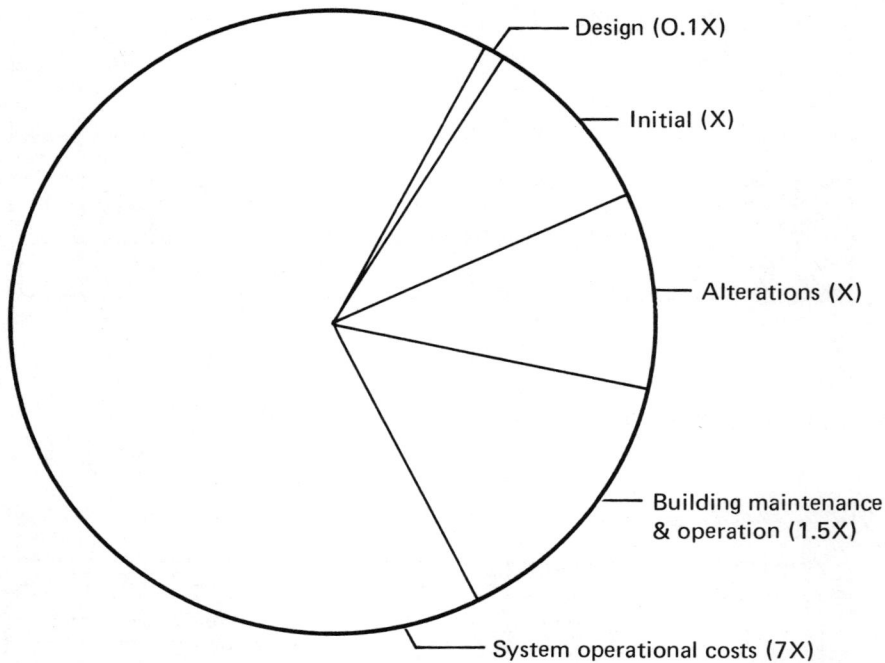

FIGURE 5. Life-cycle Costs (20 years) in facilities (hospital) construction. X = initial construction costs.

Figure 5 represents the blocks of incremental costs for a typical hospital and includes the cost of system operational personnel. The system operational costs (operating budgets) are portrayed at seven times the initial cost of hospital construction. For schools and other similar facilities, the system operational costs are approximately five times the initial costs. Because of these large expenditures, there is a need for a multidisciplinary approach to VE, including consultation with operations personnel, for better cost effectiveness.

In the VE methodology, cost optimization is achieved by a systematic development of alternate proposals for isolated high-cost areas. During this process a number of factors are involved. The following principal factors indicate the complexity of the problem and the effort required to arrive at meaningful decisions. The list is presented to promote thinking about the VE approach and to indicate how it can assist in arriving at more effective decisions.

Factors to be considered during proposal development are:

1. Availability of required design data
2. Initial and installation costs

3. Operational and maintenance requirements
4. Source of required material and availability
5. Prime and/or subcontractors' reaction and know-how
6. Conformance to a standard
7. If standards are not applicable, existence of sufficient data to develop standards
8. Impact on total design (e.g., use of aluminum wire for electrical circuits may necessitate specifying larger-size conduit and slab)
9. Impact on other necessary requirements (e.g., personnel, safety, fire protection, security)
10. Impact on system operational personnel
11. Impact on personnel required to use the facility (e.g., patient, student, typist)

Each of the above factors requires investigation, evaluation, and often additional input from various sources. From these, it should be apparent that VE is designed around a total concept and is not just a device to cheapen an item.

Figure 6 shows whose decision governs the expenditure of funds. It illustrates that the chief decision makers are the owner-using agency (through their requirements) and the architect-engineer.

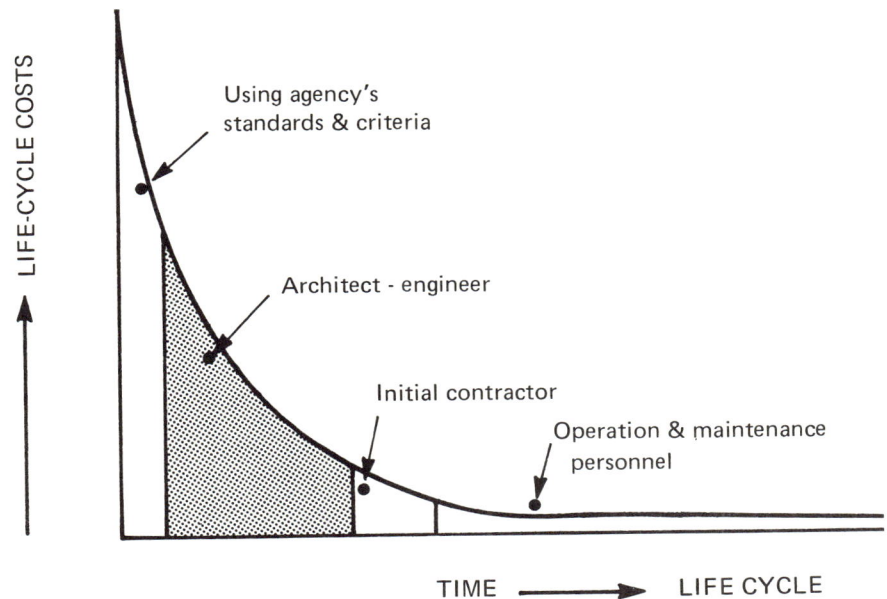

FIGURE 6. Major decision makers (persons with impact on total costs) in facility costs.

The costs involved in using agency's standards and criteria can vary considerably—from quite large when the paperwork is over 40lbs, as for one federal agency, to quite small for a private owner, who may rely completely on his architect-engineer. Generally, the architect-engineer's decisions have the greatest cost impact.

For example, the electrical engineer specifies certain power equipment which costs $4,500 per unit installed. The construction contractor during his phase can influence maintenance and operating costs slightly, and system operational personnel have to live with the functional layout provided. Figure 6 dramatically illustrates that the best place to save money is during the design phase, by critically appraising standards and criteria and the designer's major decisions. In addition, by looking at both Figures 4 and 6, an analysis of the profit levels of each participant can be made. For example, for a $1,000,000 facility, the architect-engineer's fee would approximate 5% or $50,000. His profit—if he is lucky—could be around $5,000

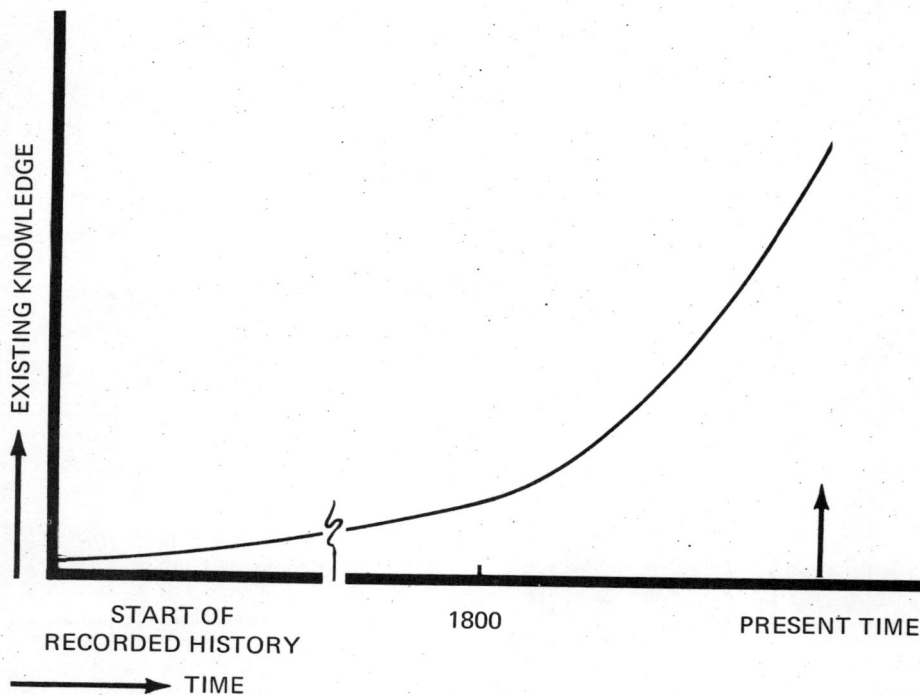

FIGURE 7. Curve of knowledge. (From W. Ogburn and M. Nimkoff. *Sociology*, 4th ed. Boston, Houghton, Mifflin, 1958).

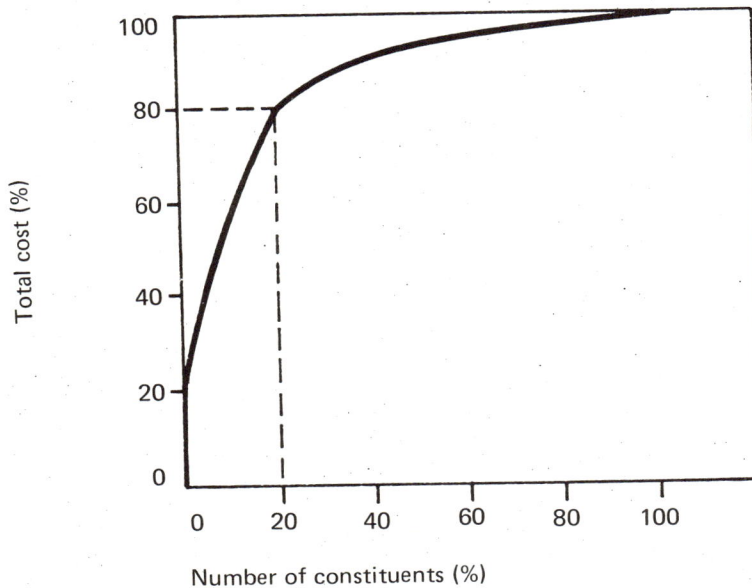

FIGURE 8.

(From *Principles and Applications of Value Engineering*, vol. 1, Department of Defense, Joint Course Book. Rock Island, Ill., U.S. Army Management Engineering Training Agency, 1967.)

for a 1-year involvement. The contractor's contract (fee) would be $1,000,000 of which his profit would approximate $50,000. Thus the chief cost decision maker's total fee is the same amount as the contractor's profit—a sad situation if the owner is really concerned about the total cost of ownership.

VE seeks to get owners to spend, in an organized fashion, a bit more during design to improve overall cost effectiveness.

There is also the technological knowledge explosion to be considered. The accumulation of knowledge shown in Figure 7 represents a forecast for the future. The curve shows that change is a basic requirement to one's life. Yet, what provisions have we made to make changes easily? Even more important, have we created a positive means to incorporate changes with significant cost impact into our operations and procedures?

Quite a few years ago, an Italian economist named Vilfredo Pareto (1848-1923), developed the curve known as Pareto's law of distribution. This curve has general application to all areas where a significant number of elements are involved (see Figure 8). It points out that in any area, a small number of elements (20%) contains the greater percentage of costs (80%). Similarly, a small number of elements will contain the greater percentage of unnecessary costs. VE uses an organized approach to isolate the elements having the greatest bulk of unnecessary costs, with the objective of developing lower-cost alternates.

Application by Government and Industry

First, with management support, the VE approach is applied in an organized full-time professional effort to high-cost areas to reduce the life-cycle costs of a facility. Savings of $5 to $20 for every $1 spent on VE have been realized in most areas.

With the use of VE, it is possible to spend $1 more in the design phase and develop cost-oriented VE changes that will save $10 or more in life-cycle costs.

Second, the VE effort is applied by a team approach utilizing creative problem-solving techniques on an organized basis. Since the lack of communication between disciplines is one of the principal causes of unnecessary costs, the use of the multidisciplinary team provides a means of overcoming this problem.

Last, incentive contracting principles are being used to inspire better, more economical decisions from designers and to enlist the support of contractors—*by paying extra fees or by rewarding them with extra profit for worthwhile proposals.*

Incentive Provisions in Contracts

The use of incentive provisions in procurement reached its peak in the Department of Defense under Secretary Robert S. McNamara. In the early sixties, as part of an effort to reduce costs, the Armed Services Procurement Regulations Committee developed contract clauses to be used for defense procurements.

The Armed Services Procurement Regulation (ASPR), Part 17, Value Engineering, sets forth contractual incentive provisions allowing successful contractors to share in any savings they can develop in their contracts. The objectives of these provisions are:

1. To take advantage of contractor know-how
2. To improve criteria
3. To reduce operation and maintenance costs
4. To provide a contractual means to share savings

Under these provisions, contractors bid on contract documents as before. The contract documents include the VE sharing provisions, but they do not affect the basis of the bid. After award, the successful contractor may submit value engineering change proposals (VECPs) to the government for changes to his contract that effect contractual savings to the government.

In fixed-price procurement contracts (including construction) the contractor submits proposals on a voluntary basis with a 50/50 sharing arrangement normally used.

For selected, negotiated procurement contracts, the government pays the contractor to engage in a VE program and shares savings at a smaller percentage, normally between 10% and 25% going to the contractor. For example, the Department of Defense included "program requirements" as part of a contract for communications facilities overseas. Here, the design and development contractor was paid an extra fee to conduct cost effectiveness studies other than those required through a normal contract for engineering service. In such studies, the contractor-designer is able to challenge government standards and criteria and to submit proposals which generate savings during the design and development of the system. The sharing arrangement for approved savings realized through this extra effort is approximately 10% for the contractor and 90% for the government.

Many questions have been asked regarding the actual impact of VECPs on parameters other than costs. Although the objective of VE is to reduce unnecessary costs without adversely affecting needed requirements, what happens in practice? To answer these questions, the American Ordnance Association made a study of 193 changes selected randomly from 2,627 implemented VECPs and determined their effect on noncost parameters. Figure 9 summarizes the results. The survey revealed that VE made significant improvements in the majority of instances, with the disadvantages quite minimal. These improved effects are called the "fringe benefits" of VE and are normally calculated by their dollar impact only in the case of readily determinable collateral savings.

From 1965 to 1971, defense department contractors using the VE incentive clauses in procurement contracts had over 6700 VECPs approved. These VECPs generated an estimated savings to the department of over $400 million dollars. The contractor averaged about $0.40 in augmented income for each dollar saved during the sharing period. The approval rate for VECPs has varied between 50% and 60%. Perhaps even more important is the trend. Estimated savings to the Department of Defense resulting from use of these provisions has increased from approximately $70 million in 1970 to over $90 million in 1971. The number of VECPs approved has reached more than 1000 annually. The number of VECPs over

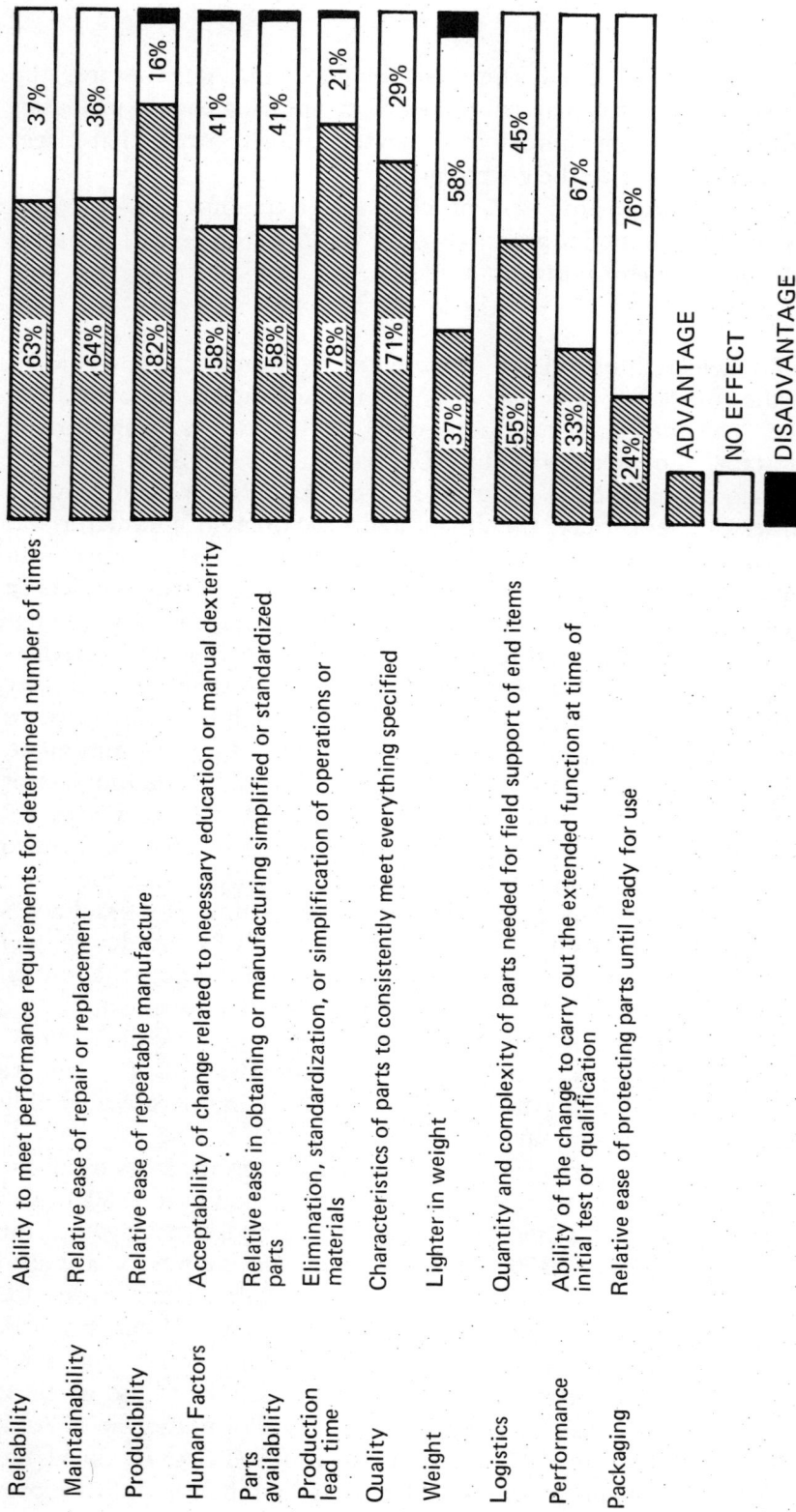

Reliability — Ability to meet performance requirements for determined number of times. 63% / 37%

Maintainability — Relative ease of repair or replacement. 64% / 36%

Producibility — Relative ease of repeatable manufacture. 82% / 16%

Human Factors — Acceptability of change related to necessary education or manual dexterity. 58% / 41%

Parts availability — Relative ease in obtaining or manufacturing simplified or standardized parts. 58% / 41%

Production lead time — Elimination, standardization, or simplification of operations or materials. 78% / 21%

Quality — Characteristics of parts to consistently meet everything specified. 71% / 29%

Weight — Lighter in weight. 37% / 58%

Logistics — Quantity and complexity of parts needed for field support of end items. 55% / 45%

Performance — Ability of the change to carry out the extended function at time of initial test or qualification. 33% / 67%

Packaging — Relative ease of protecting parts until ready for use. 24% / 76%

ADVANTAGE · NO EFFECT · DISADVANTAGE

FIGURE 9.
Total VE effectiveness. Sample of 193 implemented contractor VE changes drawn from 2627 changes. (Source: *Total Value Engineering Effectiveness.* Washington, D.C., American Ordnance Association, 1967.)

$50,000 also has increased. During 1969, for example, there were 19 approvals of over $1 million. One of these was for an estimated $22 million, with the contractor receiving $3.9 million as his share.

As for construction, from 1965 to 1971 the Corps of Engineers and the Naval Facilities Engineering Command have included incentive provisions allowing the contractor to share only in construction contract savings. They have approved over 1350 VECPs resulting in approximately $6 million in savings to the government. These savings are a small fraction of their total construction cost (0.5%), but they represent change order savings to contracts and as such, hard dollar savings to the government and extra profit to the contractor.

In 1971 the Public Buildings Service of the General Services Administration, under the impetus provided by its director, Arthur Sampson, added incentive provisions to its construction contract. In these provisions, clauses were included to allow the contractor to benefit in areas of collateral savings as well as construction contract changes. Collateral savings are reduction in operational or maintenance costs, logistic support costs, government-furnished material, etc. In the first 6 months of operation, 44 VECPs were approved with over $1 million in savings to the government. The approval rate was over 80%, indicating that the contractors conducted considerable investigation before submittal.

See Appendix A for the contract provisions used by the Corps of Engineers and the Public Buildings Service.

In addition, on March 1, 1973, the Public Building Service (Donald Parker, Value Engineering Manager) added VE Program provisions in architect-engineer and construction managers contracts for major facilities (See Appendix H). These program provisions require designers to conduct a formal VE effort outside of the standard design scope and receive extra compensation for these efforts. This action is a major step forward in reducing building costs for the General Services Administration.

As for the overall program results, the Department of Defense in fiscal year 1970 spent approximately $16 million in support of the VE program. This resulted in over $300 million saved through in-house programs and a saving of over $70 million through contractors' participation in VE provisions of procurement contracts. In other words, the results have proved that the VE program is one of the few, if not the only program in government that reduces unnecessary costs without lowering quality or performance. As a result of contract reductions through use of incentive provisions, the Department of Defense had realized in 1970:

$$\frac{\$70 \text{ million (VECP savings)}}{\$16 \text{ million (cost of program)}} = 4+ \text{ return on investment}$$

The above return does not consider the more than $300 million estimated as a return from the in-house program. This amount can be considered a fringe benefit.

As for construction, for fiscal 1971 the Corps of Engineers approved 285 VECPs of the 391 submitted. They had 30 engineers working fulltime evaluating change proposals. Hard savings from VECPs were $1,570,000:

$$\frac{\$1,570,000 \text{ (VECP savings)}}{30 \times \$20,000 \text{ (cost of program)}} = 2.6+ \text{ return on investment}$$

The above result does not include over $40 million estimated as a return from the in-house program. With these results it is amazing that more organizations, especially profit-oriented ones, do not adopt VE incentive provisions for their procurements.

Figures 10 and 11 are examples of VECPs submitted by construction contractors and approved for savings by the government.

BEFORE

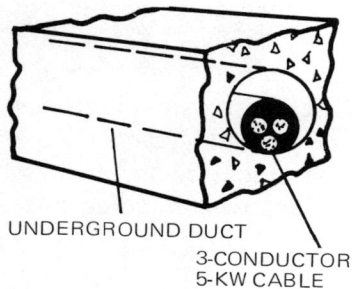

UNDERGROUND DUCT
3-CONDUCTOR
5-KW CABLE

AFTER

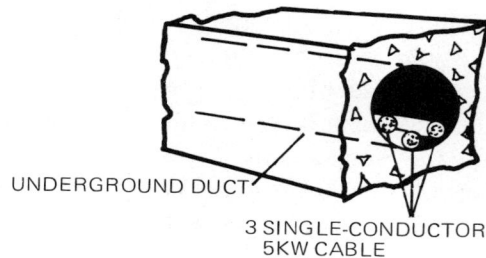

UNDERGROUND DUCT

3 SINGLE-CONDUCTOR
5KW CABLE

VE CHANGE: Change 3-conductor 5-kW cable to 3 single-conductor 5-kW cables. Installation labor exceeded extra costs of 3 single conductors. Principal savings resulted from elimination of expensive supplies required for 3-conductor cable.

SAVINGS: $5780

FIGURE 10. VE change proposal for underground cable.

BEFORE

CONSTRUCTION MATERIALS

POURED-IN-PLACE

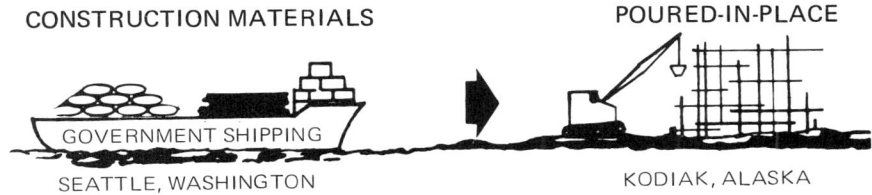

GOVERNMENT SHIPPING

SEATTLE, WASHINGTON

KODIAK, ALASKA

AFTER

PRECAST ITEMS

PRECAST UNITS

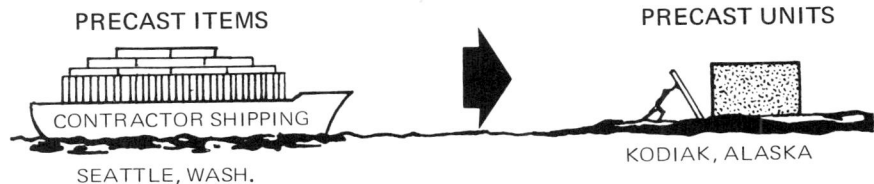

CONTRACTOR SHIPPING

SEATTLE, WASH.

KODIAK, ALASKA

VE CHANGE:

Contract called for government shipment of materials for poured-in-place construction of basement foundation walls, whereas contractor submitted VECP to precast units at Seattle for shipment at his expense. Change resulted in savings of government shippage costs; contractor's share $2000

SAVINGS: $12,980

FIGURE 11. VE change proposal for family housing project.

Application to Design and Construction

In the present method of designing a facility, the designer develops plans and specifications which conform to the criteria of the using agency. He must determine which equipment and methods are most suitable from the standpoint of economy, function, and maintenance, but within the standards and criteria set by the owner. Generally, each selection is done by an engineer or architect working on a particular aspect of the design. For example, the electrical engineer selects the generators and material for conductors, conduits, panel boxes, etc. The civil engineer selects the sewage and water systems, etc.

In some cases economic studies are conducted, such as for site selection, fuel selection, and structural system. However, in most instances, any selections or studies are made by an individual or a group within the same discipline. In some cases, a team is called together, but normally no formal job plan is followed nor are any employees assigned full-time to organize and coordinate activities or follow through on any new ideas generated.

What occurs is that each discipline generates requirements, reviews these requirements, establishes and modifies its particular criteria, and even modifies the standards and criteria of the owner. This approach may not lead to decisions which are most economical for the end function of the facility. Instead, it encourages economical decisions within each area with maximum safety factors deemed necessary by each discipline. Although this system is not totally without merit, it tends to sacrifice overall system performance in maximizing subsystem performance. The result is that total life-cycle costs are not adequately considered. Figure 12 represents how the decisions of various disciplines affect total building costs. The figure indicates that each discipline's decision can affect costs in other areas. It is in the overlapping area that VE has the greatest saving potential. It emphasizes the need of a team approach. For example, the architect, in an effort to optimize costs in his area can adversely affect the cost areas of all other disciplines. Therefore, a team approach is required for cost effective decisions in major design areas.

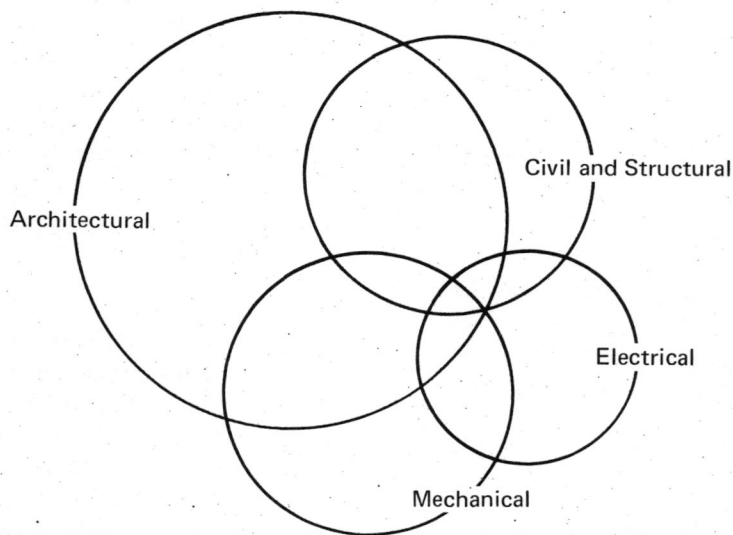

FIGURE 12.
Effect of decisions by various disciplines on total building costs. Chart is based on typical commercial high-rise office building; effect will vary according to type of structure.

How can the fresh approach outlined in the VE methodology be applied to the design of facilities?

First, it is the responsibility of management to become familiar with these new concepts. Management should establish a full-time VE effort in their own organization. The VE effort should start with a comprehensive training program for all personnel who make decisions affecting costs. The magnitude of the effort need not be large, but to succeed it must have top management support. The program should be established at a management level where it can effectively challenge design criteria, including established standards, and have access to operational and maintenance costs. It should be concerned solely with seeking out high-cost or poor-value areas and developing lower-cost alternates. This plan would supplement present methods and provide better information on which to make design selections.

Admittedly, this approach will require changes in organizational structure. But in view of the challenges of the technological explosion confronting the construction industry, changes will have to be made anyway, and now is a good time to incorporate a VE program as part of these necessary changes.

Second, the principles of incentive contracting should be used wherever feasible. For example, consideration should be given to voluntary incentive provisions for competitively bid, fixed-price contracts. By including these provisions, contractors will have the means to share in any cost-saving proposals which they submit after contract award. Funded program requirements for large, negotiated design and/or construction management contracts should also be considered. With funds available, designers and construction managers would be able to conduct additional studies not within today's fee structure on any key design selection which has a significant impact on total costs. They would also have funds to challenge an owner's criteria or specifications which in their judgment represented poor value.

The program requirement provides a means by which the design and/or management costs can be increased by a small percentage to gain a large percentage reduction in total costs. To date, program requirements for hardware or systems design-oriented contracts have ranged from 0.3% to 0.5% of total costs. The savings (target goal) projected for this effort is a 1% to 5% reduction in total costs. For example, for a $10million cost-plus-fixed-fee contract, a program requirement would range from $30,000 to $50,000 with a target savings of from $100,000 to $500,000. To date, the results from limited efforts in the construction area have generally followed these parameters. For construction management (CM) contracts, the VE efforts should range between 5% to 10% of total fees. For these efforts, documented savings should approach the total fee of the CM. Experience of one of the largest CM firms has proven this potential. See Appendix J for recommended VE requirements for CM contracts.

VE is a proved tool for cost reduction. In fact, in 1970, over 300 full-time value engineers were working in the Department of Defense. Other government agencies, such as the General Services Administration, Facilities Engineering and Construction Agency of the Department of Health, Education, and Welfare, Federal Aviation Agency, National Aeronautics and Space Administration, Veterans Administration, and various state governments (the first being Massachusetts), have established VE programs. In the private sector, many major manufacturers have adopted full-time programs. Foremost among these are Minneapolis-Honeywell, Joy Manufacturing, General Electric, Philco-Ford, Lockheed, Sylvania, Radio Corporation of America, and Thompson-Ramo-Wooldridge.

In addition, the Society of American Value Engineers (SAVE)* was formed in the late fifties to promote the principles of VE methodology.

The unprecedented challenge now confronting us requires new thinking to reduce unnecessary costs. Ask yourself the question: If I receive 10% of any savings I generated during the design phase could I turn out more economical designs? If the answer is yes, a formal VE program should be implemented.

*National Business Office, 2550 Hargrove Drive, L-205, Smyrna, Georgia 30080.

VE Job Plan

After selection of high-cost areas for study, the VE job plan is applied. Selection of high-cost areas is discussed later, under "Techniques of Project Selection." The first step after selection is to form multidisciplinary teams representing a cross section of technical fields.

This multidisciplinary approach is one of the keys to a successful study. The reason for its importance is that more and better ideas tend to be generated, greater consideration is given to the total impact of decisions on both the facility and costs, and improved communications are developed among disciplines. Engineers working on a design often are so close to the essential functions that they fail to detect areas of unnecessary costs. Each team should have one member from the major discipline that is related to the project under study to assist in obtaining pertinent information. Other team members should be knowledgeable about the system or be from other fields with no previous association with the design being studied. It is interesting to note that many new developments in materials and methods in a given field (such as space or ship construction) can be successfully applied to a design in the facilities area.

The team concentrates on determining and evaluating basic vs. secondary functions. They attempt to reduce costs without sacrificing essential *functions* instead of merely using cheaper materials, smaller quantity, or lower quality.

Each team analyzes its project, utilizing the VE job plan, which is function-centered. Special worksheets are used to aid in organizing the steps in the job plan. Figures 13 and 14 outline the significant steps and methodology of the job plan. The principal phases of the job plan are information, speculative, analytical, and proposal. An in-depth discussion of these phases follows.

Information Phase

During the information-gathering certain questions must be answered.

1. What is the item?
2. What does it do? (define the function)
3. What is the worth of the function?

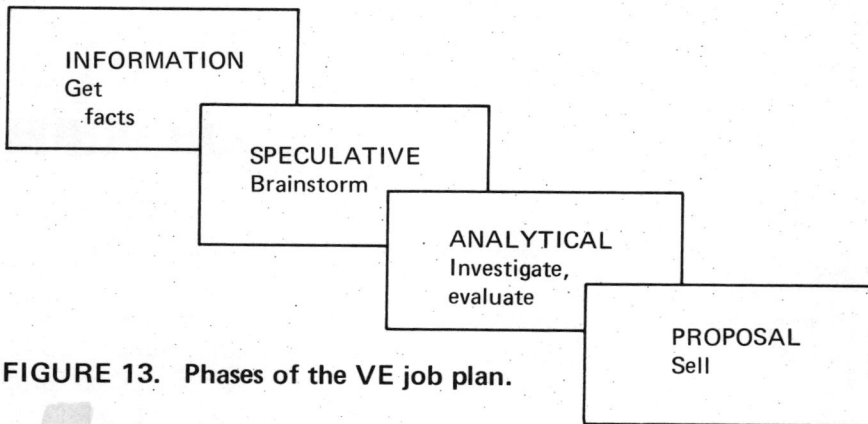

FIGURE 13. Phases of the VE job plan.

4. What does it cost?
5. What is the cost/worth ratio?
6. What are the needed requirements?
7. What high-cost or poor-value areas are indicated?

Considerable effort, ingenuity, and investigation are required to answer these questions. The group must determine what criteria and constraints existed at the time of the original design, and whether they still apply at the present time.

Other important questions considered during this phase include:

1. How long has this design been used? The age of the design is significant since it could indicate whether substitution of newly developed materials and methods should be considered.

2. What alternate systems, materials or methods were considered during the original concept? These data should be carefully weighed because they might lead us astray if the information is too old. For example, a system which might rightly have been rejected several years ago because of technical difficulties, cost, or other deficiencies, might today be able to provide the optimum requirements.

3. What special problems were or are now unique to this system? For example, in the case of an electrical distribution system, if the reasons why an above-ground system was chosen were a highly corrosive fill material and high ground water, there would be little sense in considering an underground system.

4. What is the total use or repetitive use of this design each year? What is the contemplated procurement and its cost? This cost and potential estimated savings must be weighed against the estimated cost of the VE study. If the potential savings do not exceed the cost of the study by several times, the study project should be dropped.

Objectives	Job Plan	Key Techniques	Supporting techniques	VE Questions
Obtain background	Information phase	Get all facts Determine cost	Obtain all information Work on specifics	What is it? What does it cost?
Define functions		Define function Put dollar value on specifications and requirements Determine worth	Divide problem into functional areas	What is its function? What is the value of the function?
Create ideas	Speculative phase	Blast and create	Create Innovate Defer judgment	What else will perform the function?

Evaluate basic function	Analytical phase	Evaluate:
		  Basic function
		  By comparison
			Evaluate functional areas	What ideas will perform the function?

Evaluate new ideas		Put dollar value on ideas and refine	Analyze cost, use good judgment	

Consult		Investigation
		  Suppliers
		  Companies
		  Consultants
			Investigate advanced techniques	What else will do the job?

Compare		Use Standards
		Compare:
		  Methods
		  Products
		  Materials
			Develop new ideas	

Develop alternates		Determine costs	Use teamwork	What will alternates cost?
List best ideas	Proposal phase	Extract data	Use good human relations	
Summarize		Motivate positive action	Finalize solutions	
Document			Document and present solutions for action	

FIGURE 14. Methodology of the VE job plan.

```
┌─────────────────────────────────────────┐
│                                          │
│  PROJECT_____ │
│                                          │
│  ITEM _____ PROJECT NO._____ │
│                                          │
│  BASIC FUNCTION _____ DATE _____ │
│                                          │
│  DESIGN CRITERIA:                        │
│                                          │
│                                          │
│                                          │
│                                          │
│                                          │
│                                          │
│  DESIGN HISTORY & BACKGROUND:            │
│                                          │
│                                          │
│                                          │
│                                          │
│                                          │
│                                          │
│  TEAM MEMBERS        _____ │
│                                          │
│  _____    _____ │
│                                          │
│  _____    _____ │
│                                          │
└─────────────────────────────────────────┘
```

FIGURE 15.   Form for information gathering by VE team.

Figure 15 provides a format for gathering this type of information.

In the information phase the most important steps are (1) *determining the basic and secondary functions of the items in the design* and (2), *relating these functions to cost and worth*. The functions should be stated in terms that accurately define the problem and at the same time are broad enough to generate the greatest number of alternate solutions. The *creative approach* in VE begins here because if the functions are stated too narrowly or too restrictively, creative thinking may be inhibited. In simple terms, functions are those performance characteristics which an item possesses. A design normally has both basic and secondary functions. The basic function is the primary purpose of the design. It is that function which must remain to do the job. Conceivably, all others might be eliminated, but this could not. Secondary functions are not required for their own sake—they only augment the basic function. If the design can be changed, the need for the secondary functions may be modified or even eliminated.

To facilitate the evaluation, the function of any item or design must be defined literally by two words: a verb and a noun. Longer definitions are not concise enough for VE purposes. For example, the basic function of a chair or bench is to "support weight"—*support* is the verb, *weight* the noun. The basic function of a door is to "provide access." The function of a wire is to "conduct current"; that of an elevator, to "move weight."

- The use of two words helps to avoid combining different functions and ensures that only one function will be defined at one time.
- The use of two words facilitates the task of distinguishing between primary and secondary functions because it helps to identify each function as specifically as possible.

For example, in analyzing the supporting system for a suspended pipe, the function can be defined as to "hold pipe." However, defining the function as to "restrain movement" permits an expansion of our thinking and allows a broader consideration of devices and methods. This approach accomplishes two things:

1. It allows looking into devices in unrelated fields for the solution of the problem.
2. It tends to lead to solutions which can be applied to problems in totally unrelated areas.

Suppose, for example, there is an area that is infested with mice; mousetraps have proved inadequate to cope with the problem. The conventional solution might be "to build a better mousetrap." However, a more creative way of stating the problem, such as "eliminate rodents," could lead to solutions that exclude mice from entering as well as to methods of killing them.

The next step is to determine the worth of the basic function or "value standard." Worth is defined as the lowest cost to perform the basic function in the most elementary manner feasible, within the state of our present technology. No worth is assigned to secondary functions. Worth is used as an indicator of value in the performance of a particular function. Extreme accuracy in determining this cost is not important since it is merely used for comparison. For example, the basic function of a pencil is to "make marks." This function can be accomplished by a piece of lead costing $0.02, which is about as elementary a method of making marks as possible. Therefore, the worth of making marks is $0.02.

The final step in the information phase is to determine what we are paying for this function in relation to the simplest method or design that will achieve the same results. This relationship is called the cost/worth ratio. If it is greater than two or three, poor value and high costs are indicated. The cost/worth ratio gives an indication of the efficiency of the design or item.

Figure 16 is a functional analysis of a wooden pencil. After the basic function of the pencil (to "make marks") is identified, the components of the pencil are listed in the appropriate column. Then, the function of each component is determined by the verb-noun technique. For example, the eraser has the function to "remove marks." Note that one component may have more than one function, e.g., the paint, which "protects wood" also "provides beauty." Next, the functions are evaluated to determine whether they are primary or secondary. If the function is included solely because of the method of design, it is secondary. The only component functions classified as primary are those without which the design or item *cannot* achieve its basic function. For example, all the pencil components perform secondary functions except the graphite. The remaining components are in the design because of the particular method the designer chose to "make marks."

After functions have been evaluated, the cost for each component and the total cost of the design or item is determined—in this case, $0.07. This value is then compared with the worth. As stated earlier, worth is assigned only to those components having basic function. Those items performing secondary functions have no *worth*. Finally, the cost/worth ratio is determined. The pencil has a ratio of 3.5, which would normally indicate that savings could be realized and still allow the basic function to be met.

It can be noted that the approach helps to isolate those areas in a design or product that contain high costs or represent poor value. For example, the analysis illustrates that the wood, which performs a secondary function, represents approximately one-third of the costs, and is therefore the best area to consider for potential savings.

In actual practice, on a construction project the components of a design are evaluated or compared with historical costs and team experience. Those components which appear out of line with standards are selected for in-depth study. For example, in one recent study, a water distribution system had the highest cost/worth ratio and was selected for in-depth analysis. Figure 17 is the functional analysis of the water systems of a communications facility. From this evaluation, the lawn sprinkler system (cost/worth = 10.5) and the storm drainage system (cost/worth = 7.5) were isolated for further study. Figure 18 is the functional analysis worksheet for the lawn sprinkler system. From the analysis and subsequent study, an alternate scheme was developed which reduced costs of the lawn sprinkler system by 30% (see Example 18, page 99.)

FIGURE 16. Functional analysis of a wooden pencil.

1. ITEM IDENTIFICATION						
ITEM NAME: Pencil				DATE		
				PART (OR DWG) NO.		
	FUNCTION					
COMPONENT	VERB	NOUN	KIND	REMARKS	WORTH	ORIGINAL COST
Pencil	Make	Marks	B			$0.07
Eraser	Remove	Marks	S			0.01
Ferrule	Hold	Eraser	S			0.005
Wood	Hold	Lead	S			0.025
Paint	Protect	Wood	S			0.005
	Provide	Beauty	S			
Markings	Identify	Product	S			0.005
Graphite	Makes	Marks	B		0.02	0.02
Carbon				Cost/Worth Ratio=		
				7/2= 3.5		

FIGURE 17. Functional analysis of water systems.

PROJECT ___Capitol Hill___

BASIC FUNCTION ___Supply Water___

ITEM ___Water Distribution System___

DATE ___July 11, 1972___

QTY.	UNIT	COMPONENT	FUNCTION VERB	FUNCTION NOUN	KIND	EXPLANATION	*WORTH	ORIGINAL COST
	1	Lawn sprinkler system	Supply	Water	B	Complicated system	$2,000	$21,000
						C/W = 21/2 = 10.5		
	2	Fire protection system	Supply	Water	B	C/W =18.4/8 = 2.3	8,000	18,400
	3	Domestic water supply	Supply	Water	B	C/W =15/8 =1.9	8,000	15,000
						Extend 6in. supply line to bldg. then		
						meter and fire line		(1,500)
	4	Sanitary waste removal	Remove	Water	B	C/W =27/17.5 =1.5	17,500	27,000
	5	Storm drainage system	Remove	Water	B	C/W =15/2 =7.5	2,000	15,000
						TOTAL	$37,500	$96,500

* (1)Hose bib and hose, 6 each. (2)Water to each floor w/hose bib and hose. (3)Water to each floor. (4)Reduce number of fixtures, roughly 1/2. (5)Let water out through scuppers, 2 each side.

FIGURE 18. Functional analysis of lawn sprinkler system.

PROJECT   Capitol Hill

BASIC FUNCTION   Water Vegetation

ITEM   Lawn Sprinkler System

DATE   July 11, 1972

QTY.	UNIT	COMPONENT	FUNCTION VERB	FUNCTION NOUN	FUNCTION KIND	EXPLANATION	WORTH *	ORIGINAL COST
		Pipe and fittings	Carry	Water	B		$2,000	$9,000
		Valves and controls	Control	Flow	S		N/A	4,350
		Nozzles	Spread	Water	S		N/A	6,850
		Bubblers	Spread	Water	S		N/A	700
		Backflow preventer	Prevent	Contam- ination	S		N/A	100
		Water	Sustain	Growth	B		0	
						TOTAL	$2,000	$21,000
						Cost/Worth Ratio= 21,000/2,000=10.5		

*Worth based on approximately 6-each hose bib and garden hoses.

The function-cost-worth approach outlined above is the key difference between VE and other cost-reduction techniques. Where the organized approach is used more and better results are invariably achieved.

There are other useful methods of performing a functional analysis. For example, Figure 19 is a graphic presentation of the functional analysis technique applied to an electrical distribution system. The system was selected for review since the cost of

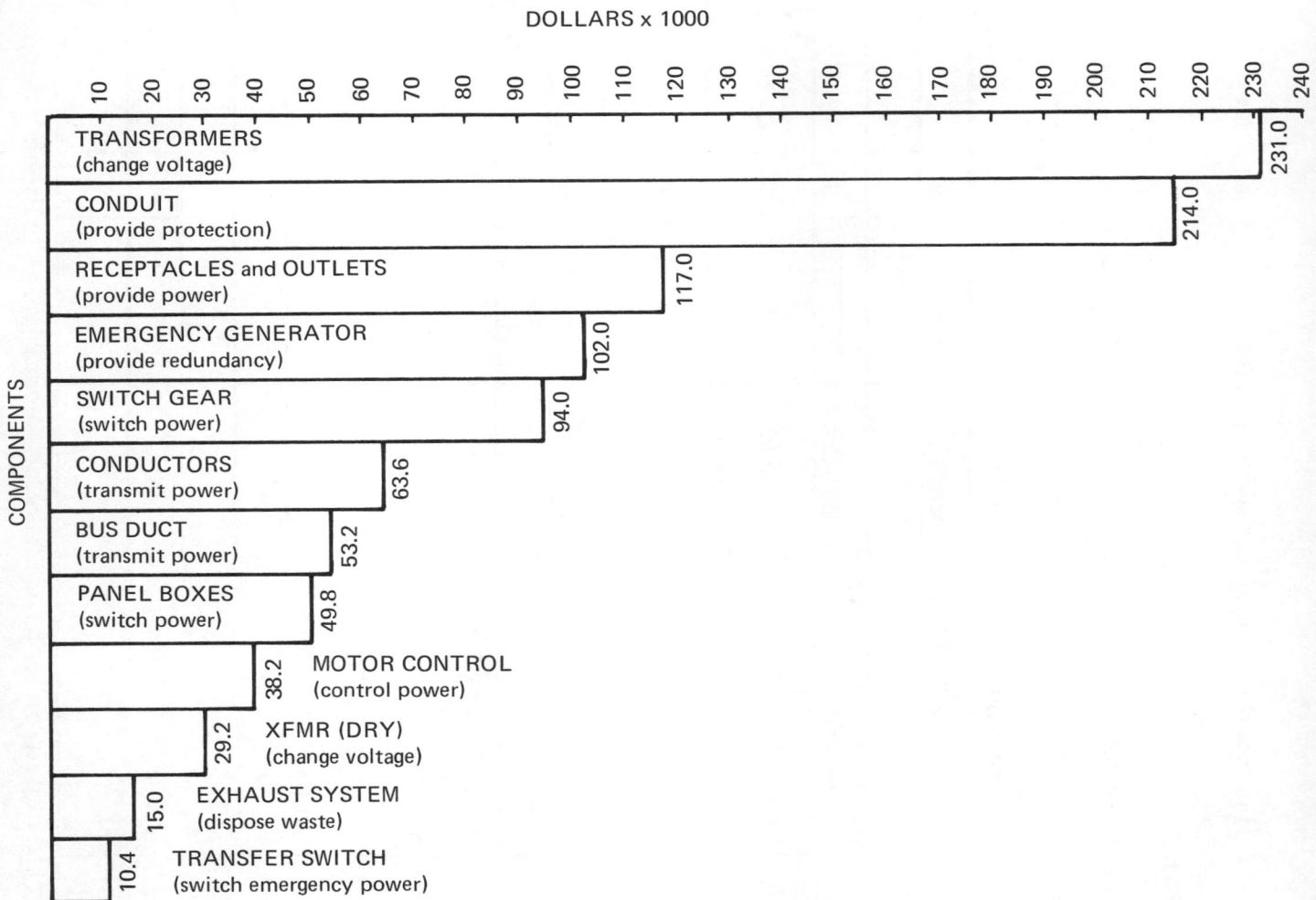

DOLLARS x 1000

COMPONENTS

TRANSFORMERS (change voltage)	231.0
CONDUIT (provide protection)	214.0
RECEPTACLES and OUTLETS (provide power)	117.0
EMERGENCY GENERATOR (provide redundancy)	102.0
SWITCH GEAR (switch power)	94.0
CONDUCTORS (transmit power)	63.6
BUS DUCT (transmit power)	53.2
PANEL BOXES (switch power)	49.8
MOTOR CONTROL (control power)	38.2
XFMR (DRY) (change voltage)	29.2
EXHAUST SYSTEM (dispose waste)	15.0
TRANSFER SWITCH (switch emergency power)	10.4

FIGURE 19. Functional analysis of hospital electrical distribution system, total cost $1,017,000.

$9.50/sq. ft. appeared excessive to the VE team. The worksheet points out quite strongly the high costs of the transformers and conduits. Why should these areas of secondary function contain approximately 45% of the total costs, and cost twice as much as the conductors and bus duct which perform the basic function of the electrical distribution system—"transmit power"? As for the cost/ worth ratio for the system, the only elements performing the basic function "distribute power" are the conductors and bus duct ($117,000). All other elements are performing secondary functions. If we take the total cost and the worth value, the cost/worth ratio is approximately $1,000,000/$117,000 or 8.5. Poor value is definitely indicated.

The follow-up VE study revealed that the designer had used a back-up transformer for each secondary transformer throughout the building. Although this system resulted in an optimum electrical design, it put the building cost out of budget. A compromise design was developed which, although not as reliable as the original, met all needed user requirements at a cost he could afford.

Figures 20 and 21 show this technique applied to a heating, ventilation, and air-conditioning (HVAC) system. The system was selected for study because the cost per square foot of the design compared unfavorably to the cost per square foot of similar buildings

Lights	210 tons	= 48.0%
Fresh air	104 tons	= 23.9%
People	58 tons	= 13.3%
Roof	43 tons	= 9.8%
Appliances	10 tons	= 2.3%
Partitions	8 tons	= 1.8%
Wall	3 tons	= 0.7%
Door	1 ton	= 0.2%
Total	437 tons	= 100.0%

Basic function: Provide comfort (people)

FIGURE 20. Functional analysis of air-conditioning load distribution (tons of refrigeration).

Dollars x 1000

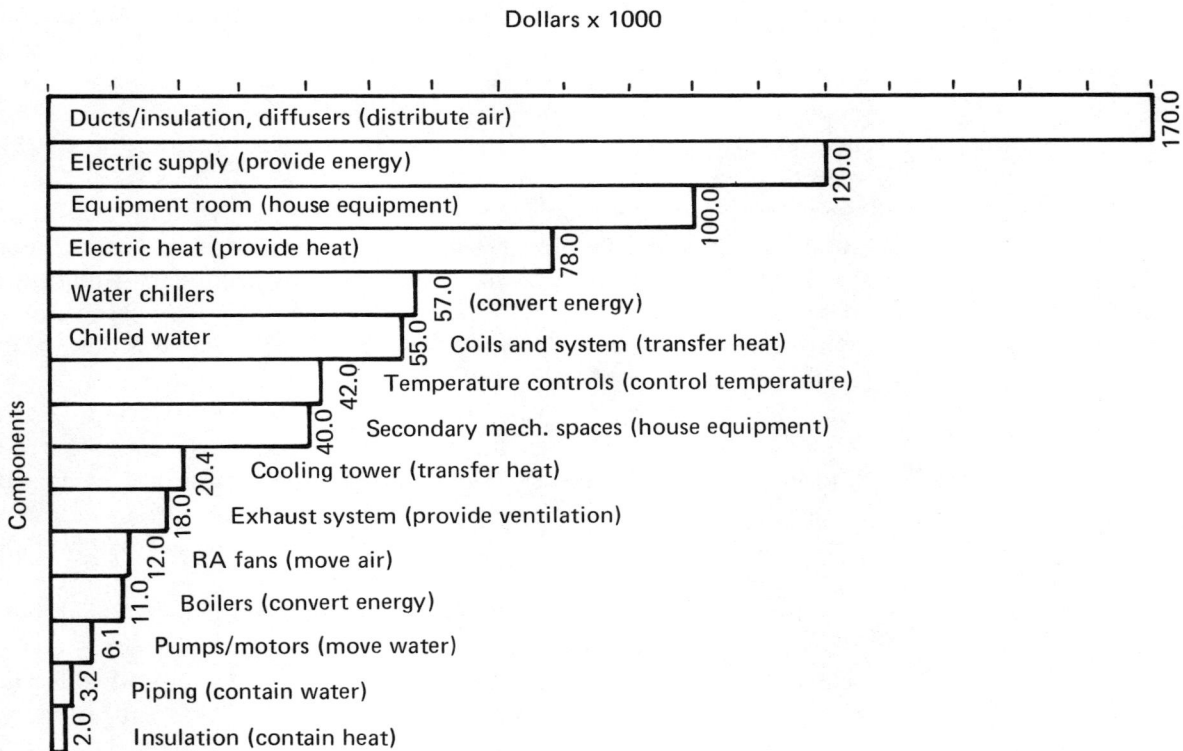

FIGURE 21. Functional analysis of air-conditioning system, total cost $734,700.

being constructed by competitors. The team decided first to functionally analyze the system loads. Figure 20 points out that the heat load from the lights is inordinately high. Lighting required 48% of the total load and, since it is a secondary load area, it represented the best area to review for cost savings. In fact, of the total load of 437 tons, only 104 tons for fresh air and 58 tons for people load are primary load areas. The team felt that significant cost saving could be achieved by a critical study of the system.

Figure 22 indicates the reason for high costs. The original design had numerous 300-W incandescent fixtures throughout the facility. It was these fixtures which impacted heavily on lighting heat loads. That is, each $20 electrical fixture affected HVAC costs by more than $100. The in-depth study revealed that the original designer was meeting the requirements of the owner's sales personnel. The salespeople did not realize the cost impact of their decision, and neither the electrical designer nor the mechanical designer questioned these high-cost decisions.

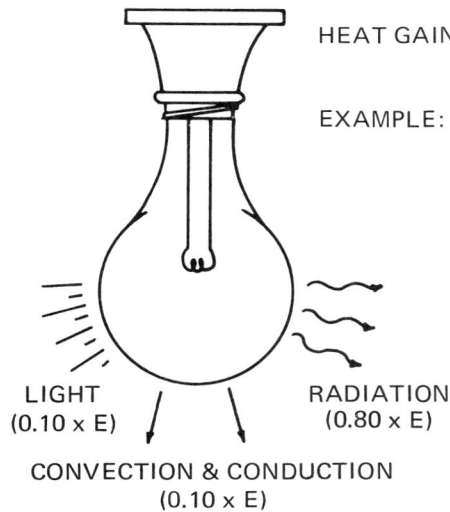

HEAT GAIN = E x 3.4 BTU/HR where E = power to light in watts

EXAMPLE: A 300-W light would generate 300 x 3.4 or 1020 BTU/HR which would require an additional 0.1 ton of refrigeration

A ton of refrigeration cost $1000 installed; thus, it costs $100 in capital investment to cool one 300-W light fixture

LIGHT
(0.10 x E)

RADIATION
(0.80 x E)

CONVECTION & CONDUCTION
(0.10 x E)

COST OF FIXTURE (installed)	$ 20 ea.
IMPACT ON AIR-CONDITIONING	$100 ea.

FIGURE 22. Impact of incandescent lighting loads on air-conditioning.

Upon review of the team study, the owner compromised and allowed localized use of incandescent and greater use of other types of lighting, which significantly reduced the light (heat) load.

A functional analysis (Figure 21) of the air-conditioning system was also conducted. It pointed out that the optimum places to reduce costs were in the duct distribution and equipment room. These items account for the bulk of costs (over 35%), with only secondary functions being performed. Of course, by reducing load, all component costs (especially electric supply) are reduced.

As for other potential solutions, one alternate would be to use more expensive electrical fixtures vented into the ceiling space and waste the heat. The extra cost of the fixtures would be more than offset by the reduction in air-conditioning requirements.

Figure 23 is a graphic functional analysis of the air-conditioning system of a typical telephone communications building. From the analyses of subsystems, the team focused on reducing the area used to house equipment rather than the mechanical equipment or duct. It was unusual, but this space represented 28% of the total costs and was the largest area of secondary costs. This example points out the

PROJECT Communications Building   ITEM HVAC SYSTEM – Actual

BASIC FUNCTION Provide Comfort   DATE November 1973

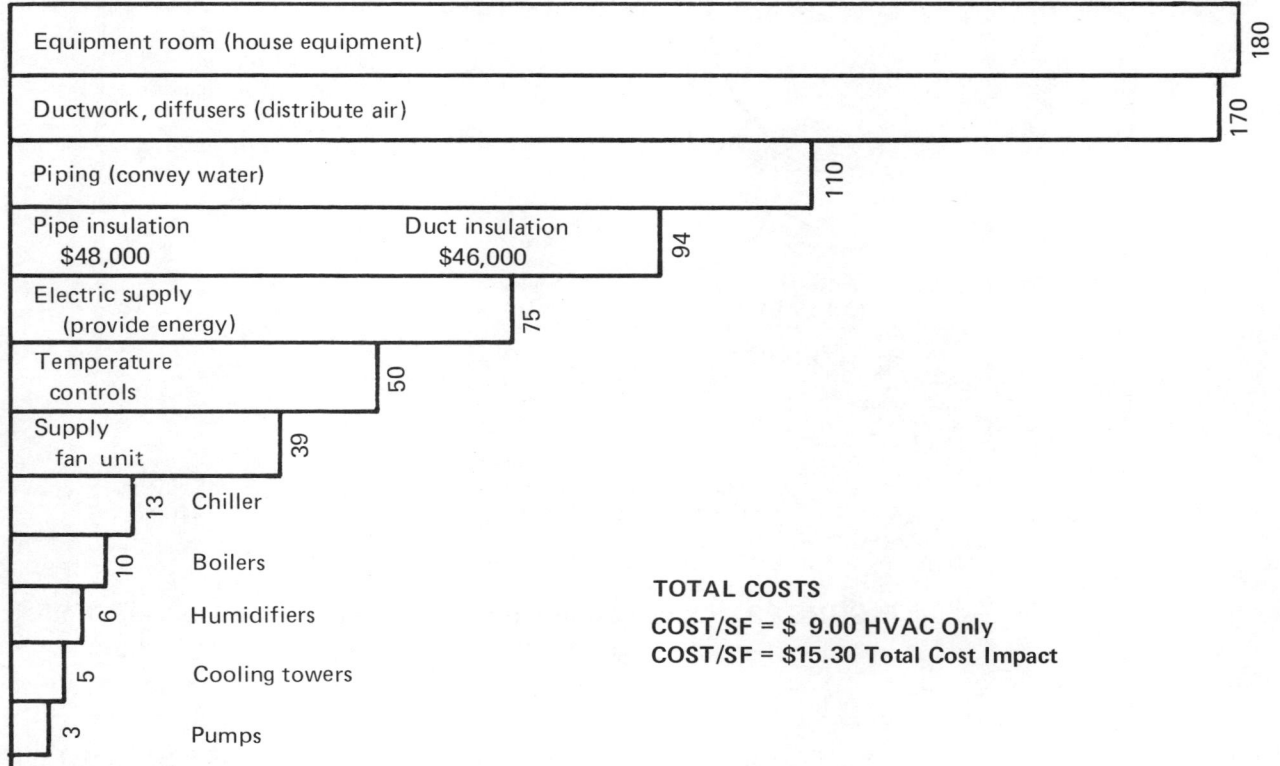

Equipment room (house equipment) — 180

Ductwork, diffusers (distribute air) — 170

Piping (convey water) — 110

Pipe insulation $48,000   Duct insulation $46,000 — 94

Electric supply (provide energy) — 75

Temperature controls — 50

Supply fan unit — 39

Chiller — 13

Boilers — 10

Humidifiers — 6

Cooling towers — 5

Pumps — 3

TOTAL COSTS
COST/SF = $ 9.00 HVAC Only
COST/SF = $15.30 Total Cost Impact

FIGURE 23. **Functional analysis of air-conditioning system in communications building.**

interdisciplinary nature of cost evaluation and the need for the team approach: In this example the architectural impact of the mechanical system was where the greatest savings potential existed.

As for a lighter side of VE, during a recent VE workshop, the group decided to apply the function-cost-worth approach to a real world problem. Does the functional analysis approach have application to problem solving in the matrimonial area? To answer this question, the group conducted a life-cycle and value analysis, using functional analysis techniques, of what a young, 20-year-old, single male has to consider when entering into a childless marriage. The group isolated "provide homekeeper" and "provide lover" as the two primary functions a young man looks for in matrimony. Figure 24 is the graph developed to aid the young man in determining cost/worth ratios over his marriage span. It may be noted that the young man has a "good deal" initially in that his direct costs are substantially less than the estimated worth. However, as the years go by, direct cost rises dramatically (similar to the labor curves of Figure 3), and after approximately 17 years the trading point occurs. After this point the usury zone is entered, and other functions besides "provide lover" or "provide housekeeper" must be considered. For the speculative type of male, the escape or itch period is plotted. It is during this period that the male has received greatest worth for the direct costs expended.

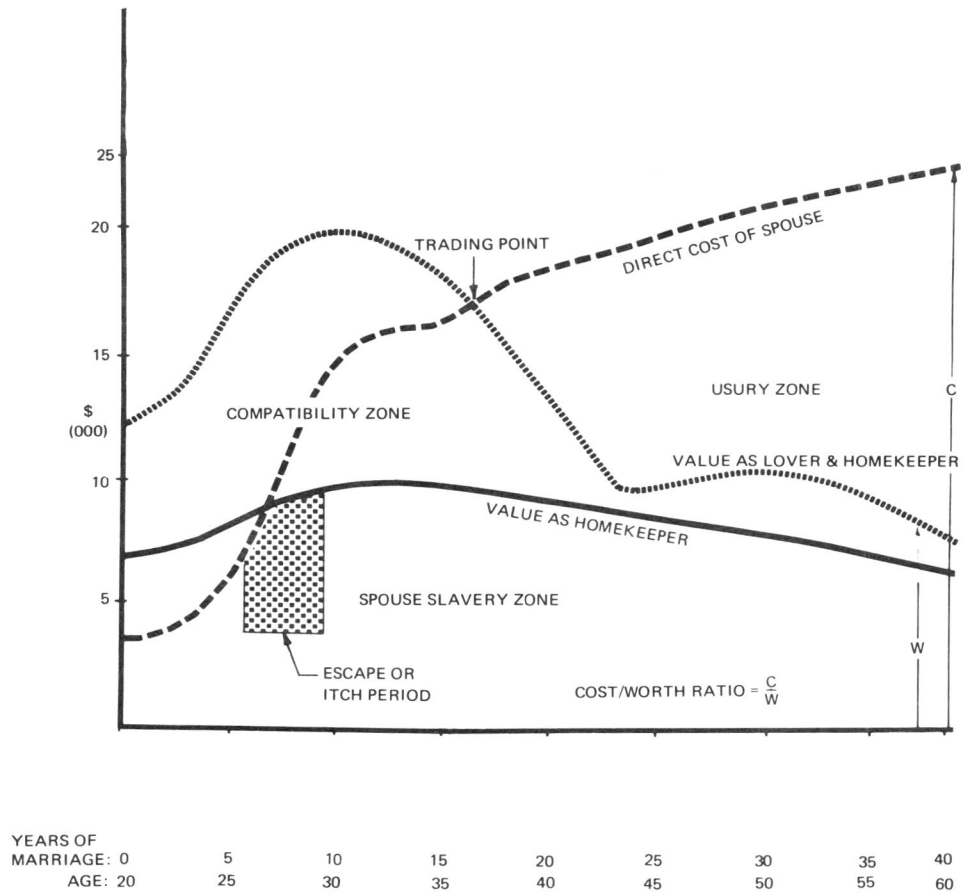

FIGURE 24.   Life cycle and value analysis of a childless marriage.

Speculative Phase

During this phase of the job plan the principal question to be answered is: in what alternate ways can the necessary function be performed? This phase is designed to introduce new ideas to perform the basic function. Therefore, it is necessary to fully understand the problem and, by using problem-solving or creative techniques, to generate a number of ideas which introduce lower-cost alternates. These additional ideas not only increase the opportunity for cost savings, but also enhance optimum solutions for design problems.

The creativity process starts early in life for every individual and maximizes at 4 or 5 years of age (Figure 25). From then on, creative thinking is restricted by parental and school controls and social and legal requirements until many lose their inherent ability to be creative. In fact, one can lose creativity by narrowing his environment as shown in Figure 25.

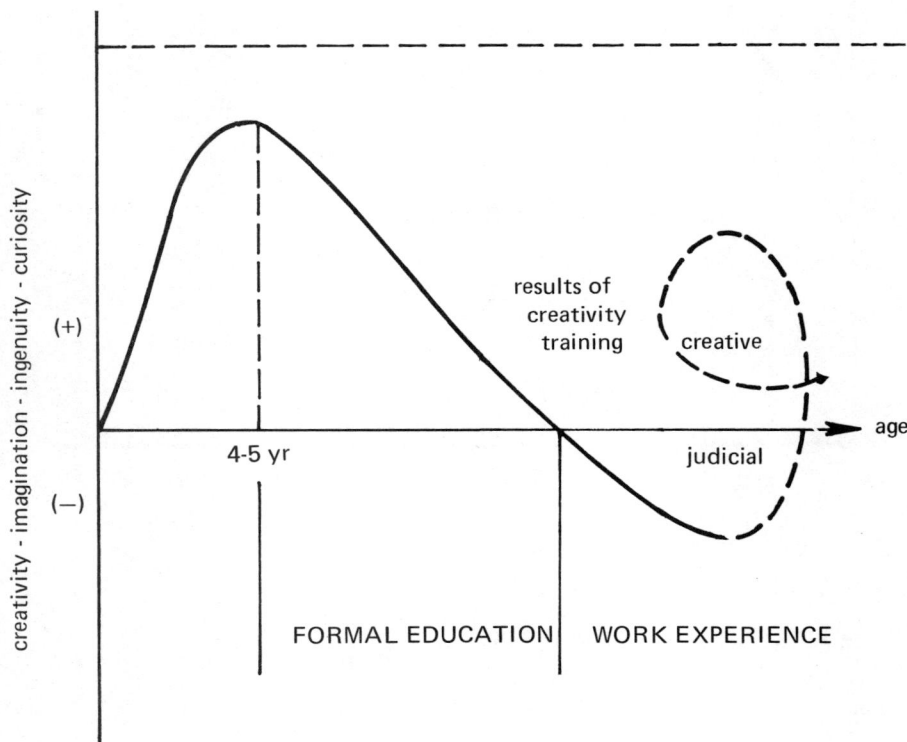

FIGURE 25. Relation of age to creativity.

What can be done: For one thing, it must be recognized that the creative process requires an alert mental attitude and a responsiveness to change. Also, it must be recognized that regeneration of creativity can definitely be advanced through training. This phenomenon is indicated by the loop on the curve. By training and practice, creative ability can be improved by learning to recognize and overcome barriers to creativity. In addition, one can broaden his experience by other means, such as adopting a creative hobby.

One aspect of training is very important. Emotional blocks, such as the fear of making a mistake or of appearing foolish, must be removed in order to maintain a positive approach to creativity.

Two major creativity approaches, classified under free-association techniques, are brainstorming and the Gordon technique.

Brainstorming

The foremost approach to creativity in VE is the brainstorming technique. A brainstorming session is a problem-solving conference wherein each participant's thinking is stimulated by others in the group. The typical brainstorming session consists of four to six people of different disciplines sitting around a table and spontaneously producing ideas related to the performance of the required function. During the session the group is encouraged to generate the maximum number of ideas. No idea is criticized. Judicial and negative thinking is not permitted.

The quantitative result of a multidisciplinary group in generating ideas has no parallel. The team concept not only results in a large number of ideas but also improves the creative ability of the participants. There are many reasons for the high quantity of group ideas obtained, but perhaps the most important one is the aspect of interdisciplinary communication. Many times, one member's idea motivates the associative processes of the other group members. This phenomenon produces a chain reaction, triggering many ideas; and the cycle repeats itself. Research tests conducted by the University of Buffalo demonstrated that groups generate from 65% to 93% more ideas than individuals working alone.*

Brainstorming Rules. To insure that all ideas are accepted, ground rules are established prior to opening the session. These rules are: no criticism of ideas is allowed, "free wheeling" is encouraged, a large quantity of ideas is wanted, combining and improving ideas is sought, and all ideas are recorded. Thus, the pressures that normally

Value Control, Autonetics, Anaheim, Calif., 1963, p. 42.

exist in an ordinary conference are removed. Since time is not devoted to criticism or evaluation, a great many more ideas per unit of time are generated. This approach is known as the principle of "deferred judgment."

Actual Application. Experience has shown that engineers in particular, and architects to a lesser extent, find it difficult to participate in brainstorming sessions. It is necessary to continually point out that any evaluation or criticism must be withheld until a later phase. Only after a number of sessions is free wheeling achieved.

The Gordon Technique *

The other type of free-association technique closely related to brainstorming is the Gordon technique. This is also a group conference method in which free-flowing idea discussion is encouraged. Unlike the brainstorming technique, however, in the Gordon approach only the group leader knows the exact nature of the problem. The leader asks the group questions which lead the team into the generation of ideas.

Whatever creative thinking technique is used, the objective is to arrive at a number of alternate ways for performing the function. The team lists the ideas on a Creative Idea Listing worksheet shown in Figure 26. An example of a creative idea listing is shown in Figure 27.

Analytical Phase

In this phase, sometimes called the evaluation and investigation phase, the team examines, and then develops the alternates generated during the preceding phase into lower-cost alternate solutions. The principal tasks are to evaluate, refine, and cost analyze the ideas and to list feasible alternates in order of descending savings potential.

During this phase the ideas must be refined to meet the necessary environmental and operating conditions of the particular situation. Ideas which obviously do not meet these requirements are dropped. Ideas having potential, but beyond the capability of our present technology, are put aside for possible discussion with progressive manufacturers. The remaining ideas are potentially workable and are cost analyzed. Those showing worthwhile savings are then listed along with their potential advantages and disadvantages. Advantages might be light weight, reusability, low cost. Disadvantages might be high maintenance, excessive construction time, too many pieces. Ideas whose advantages outweigh the disadvantages and which indicate the greatest cost savings are selected for further evaluation.

* *Synetics: The Development of Creative Capacity,* William J. J. Gordon, Harper & Row, New York, 1961.

PROJECT _____

ITEM _____ PROJECT NO. _____

BASIC FUNCTION _____

UNINHIBITED CREATIVITY   DATE

(Don't Evaluate Idea-------------Idea Refinement is Later)	
1.	18.
2.	19.
3.	20.
4.	21.
5.	22.
6.	23.
7.	24.
8.	25.
9.	26.
10.	27.
11.	28.
12.	29.
13.	30.
14.	31.
15.	32.
16.	33.
17.	34.

FIGURE 26.   Form for creative idea listing by VE team.

PROJECT _____ RELOCATION OF FUEL STORAGE _____

ITEM _____ FUEL PIER DECK _____ PROJECT NO. _____

BASIC FUNCTION ____ SUPPORT LOAD _____

UNINHIBITED CREATIVITY   DATE

(Don't Evaluate Idea-------------Idea Refinement is Later)

1.	PIP concrete	18.	Fiberglas-epoxy deck
2.	Timber	19.	Fill
3.	Precast deck panel	20.	Prestressed deck
4.	Wood on steel stringers	21.	Gunite
5.	Steel grate on beams	22.	Plastic wire
6.	Timber stringers under PIP concrete	23.	Underwater tunnel
7.	Waffle slabs	24.	
8.	Steel deck	25.	
9.	Wood on timber stringers	26.	
10.	Monorail	27.	
11.	Rail tracks	28.	
12.	Arches	29.	
13.	Suspension system	30.	
14.	Rope bridge	31.	
15.	Wire cables	32.	
16.	Bailey bridge	33.	
17.	Asphaltic concrete deck	34.	

FIGURE 27. Creative idea listing for relocation of fuel storage.

When the group members consider the disadvantages of each particular idea, they ask, How can we overcome the disadvantages? The group lists what information, test, approval, or action is required to make the disadvantages acceptable for this particular application. For example, a change in specification or design criteria may be required. Each team member is assigned an aspect of the problem to investigate and determine whether the disadvantages can be overcome. During this phase, manufacturers, contractors, and specialists are consulted. The objective is to determine if an alternate idea can meet needed requirements at a cost lower than the original design.

As an aid to organizing the analytical phase, the worksheet in Figure 28 was developed. The worksheet is useful in rating and selecting ideas worthy of further evaluation and investigation. Figure 29 shows the analysis of the ideas selected for a sprinkler system design.

The VE group must use all available sources of information to determine if the alternate they select is truly less costly, and performs the *required* functions without impairing the essential quality, reliability, or maintainability. An important aspect of the problem is the determination of total costs. One solution might offer a lower acquisition cost, but at the same time result in higher cost for the life of the system. That is, the initial cost might be lower, but the overall cost to the user could be higher because of increased operational or maintenance costs. VE considers the total costs involved.

Two types of worksheets are used to organize the cost information of alternates selected for evaluation. The first is for determining the initial cost of alternates; the second is for determining their total costs. Figures 30 and 31 illustrate the use of the *initial cost* worksheets for the automatic voice level control and the lawn sprinkler systems. These worksheets are typical of those in use by design and construction firms.

Usually the determination of initial costs is not difficult. However, in the area of total cost impact, greater effort is required. Figure 32 is a guide worksheet that summarizes *total costs* or life-cycle costs. Figures 33 and 34 are HVAC and Sprinkler System examples using this worksheet. Since anticipated, or future costs come into play on these worksheets, we are dealing with variables that rely upon judgment. It is vital that these variables be based upon the team's judgment.

PROJECT _____ DATE _____

ITEM _____ DRAWING NO. _____

BASIC FUNCTION _____

Ideas Selected from Prior Worksheet	Potential Advantages	Potential Disadvantages	Idea Rating
List potentially work-able ideas related to basic function	*Such as: Low cost, Simplicity, Light weight, Standard part, Mass produced, Easy to repair, Low Maintenance, Reusable*	*Such as: Not fully developed, Limited availability* *In some cases, opposites of advantage list*	*Use a rating system of 1 to 10 with 10 being the best idea*

Action Required to Develop Ideas	*Indicate what ideas are to be given further consideration.* *Check latest information from trade associations, technical societies, research reports, commercial or government specialists, user or agency requirements, etc., as means to confirm the potential of the ideas.*

TEAM MEMBERS _____ _____

_____ _____

_____ _____

FIGURE 28.   Form for organizing the analytical phase of the VE job plan.

PROJECT _____Capitol Hill_____ DATE ___July 12, 1972___

ITEM _____Lawn Sprinkler System_____ DRAWING NO. _____M-2___

BASIC FUNCTION ___Water Vegetation___

Ideas Selected From Prior Worksheet	Potential Advantages	Potential Disadvantages	Idea Rating
1. Garden hose	Cheaper. Portable. Easily replaced. Concentrate water.	Needs replacing. Requires more man-power hours. Safety hazard.	5
2. Natural rainfall	No cost. No labor. No main-tenance.	Limits amount of vegetation. High risk of losing plants. Limits type of vegetation.	2
3. Reduce amount of vegetation	Less time re-quired for maintenance. Requires less initial cost.	Tennis club would object.	8*
4. Change type and head pattern	Less initial cost. Less maintenance.	Possible dry areas.	7*
Action Required to Develop Ideas	REMARKS: *Selected for cost analysis.		

TEAM MEMBERS ___Wayne Spiller___   ___Tom Meyers___

___Ralph Phillips___   _____

___Tom O'Neal___   _____

FIGURE 29. Analysis of creative ideas for lawn sprinkler system.

| SYSTEM: Automatic voice level control
SUBSYSTEM: Limiter
UNIT: Hardware | UNIT | | TOTAL COST |
	QUANTITY	COST	
ORIGINAL Automatic level control (12 channels per unit) Total	5 each	$5,628.57	$28,142.85 $28,142.85
Alternate 1 Automatic voice level control (12 channels per unit) Total	5 each	$3,162.00	$15,810.00 $15,810.00
Alternate 2 Varistor (60 required per van) Miniature jackfield (96 Jacks) Total	60 1	$ 0.27 80.00	$16.20 80.00 $96.20

FIGURE 30. Initial cost worksheet for automatic voice level control.

| SYSTEM Water Distribution
SUBSYSTEM: Lawn Sprinkler System
UNIT: Lawn Sprinkler System | UNIT | | TOTAL COST |
	QUANTITY	COST	
ORIGINAL	1	$21,000	$21,000 (bid)
Alternate 1 (Idea 3) Reduce amount of vegetation Heads Drain box Zone control	88 heads deleted	$ 100	$ 8,800
		DIFFERENCE	$12,200
Alternate 2 (Elmer's idea) Relocate park; retain sprinkler as is, except for parking lot; remove retaining wall. *NOTE: Savings of $8100 in construction of lot.	180'0" long retaining wall/14 heads	$ 45.LF @6'0" high $140	$ 2,000
		DIFFERENCE	$19,000*
Alternate 3 (Idea 4) Change type and head pattern Reduce quantity of heads and reduce quantity of bubblers	Lump sum 35	 $100	$ 8,500 3,500
		TOTAL DIFFERENCE	$12,000 $ 9,000
Alternate 4 Eliminate 14 heads $2000 Delete portion of landscaping in parking lot and corners and add concrete/asphalt. Difference $1000	Lump sum		$ 2,000 1,000
		TOTAL DIFFERENCE	$18,000 $ 3,000

FIGURE 31. Initial cost worksheet for lawn sprinkler system. See Figures 17, 18, and 29, and Example 18, page 102, for additional information.

PROJECT_____ DATE _____

SYSTEM OR ITEM _____ PROJECT NO. _____

		ORIGINAL	ALT_____	ALT_____
INSTANT CONTRACT	**INITIAL COST IMPACT**			
	1. Base Cost			
	2. Interface Costs			
	(a)			
	(b)			
	(c)			
COLLATERAL COSTS	SUB-TOTAL INSTANT CONTRACT			
	1. Other Initial Costs			
	(a)			
	(b)			
	(c)			
	TOTAL INITIAL COST IMPACT			
REPLACEMENT COSTS	**LIFE-CYCLE EXPENDITURES**			
	(a) Year			
	(b) Year			
	(c) Year			
LIFE-CYCLE COSTS	**ANNUAL OWNING & OPERATING COSTS**			
	1. Initial Cost—Amortized @ ____ years @____% interest			
	2. Replacement Costs—Amortized @			
	(a)____ years @____%			
	(b)____ years @____%			
	(c)____ years @____%			
	3. Annual Costs			
	(a) Maintenance			
	(b) Operations			
	(c)			
	TOTAL ANNUAL OWNING & OPERATING COSTS			

FIGURE 32. Form for Life Cycle Cost analysis.

	PROJECT	Research Facility		DATE	Nov.15,1972	

PROJECT: Research Facility   DATE: Nov.15,1972
SYSTEM OR ITEM: HVAC System   PROJECT NO. 3

		ORIGINAL	ALT_____	ALT_____
INSTANT CONTRACT	**INITIAL COST IMPACT**			
	1. Base Cost	$802,500	$525,000	$738,000
	2. Interface Costs			
	(a) Electrical	120,000	100,000	160,000
	(b)			
	(c)			
COLLATERAL COSTS	**SUB-TOTAL INSTANT CONTRACT**	922,500	625,000	898,000
	1. Other Initial Costs			
	(a) Owner Supplied Equipment	64,000	64,000	60,000
	(b)			
	(c)			
	TOTAL INITIAL COST IMPACT	$986,500	$689,000	$958,000
REPLACEMENT COSTS	**LIFE-CYCLE EXPENDITURES**			
	(a) Year 12th	N/A	50,000	N/A
	(b) Year			
	(c) Year			
LIFE-CYCLE COSTS	**ANNUAL OWNING & OPERATING COSTS**			
	1. Initial Cost—Amortized @ (0.0872) 20 years @ 6 % interest	86,000	60,100	83,500
	2. Replacement Costs—Amortized @ (a) 20 years @ 6 % *	N/A	2,175	N/A
	(b) ___ years @ __ %			
	(c) ___ years @ __ %			
	3. Annual Costs (a) Maintenance	25,000	20,000	16,000
	(b) Operations	30,000	35,000	25,000
	(c)	------	-----	-----
	TOTAL ANNUAL OWNING & OPERATING COSTS	$141,000	$117,275	$124,500

*Based on present worth of $25,000

FIGURE 33. Life Cycle Cost analysis for HVAC systems.

		ORIGINAL	ALT #1	ALT #2
INSTANT CONTRACT	**INITIAL COST IMPACT**			
	1. Base Cost	$21,000	$12,000	$ 9,000
	2. Interface Costs			
	(a)			
	(b)			
	(c)			
COLLATERAL COSTS	**SUB-TOTAL INSTANT CONTRACT**			
	1. Other Initial Costs			
	(a) Redesign	-----	1,000	1,000
	(b) Logistic Support	500	300	300
	(c)			
	TOTAL INITIAL COST IMPACT	$21,500	$13,300	$10,300
REPLACEMENT COSTS	**LIFE-CYCLE EXPENDITURES**			
	(a) Year	N/A	N/A	N/A
	(b) Year			
	(c) Year			
LIFE-CYCLE COSTS	**ANNUAL OWNING & OPERATING COSTS**			
	1. Initial Cost—Amortized @ .075			
	40 years @ 7% interest	$ 1,615	$1,000	$ 775
	2. Replacement Costs—Amortized @			
	(a)____years @ ___ %	N/A	N/A	N/A
	(b)____ years @___ %			
	(c)____ years @ ___%			
	3. Annual Costs			
	(a) Maintenance Materials & Labor	400	300	250
	(b) Operations	3,500	3,000	3,000
	(c)	----	-----	-----
	TOTAL ANNUAL OWNING & OPERATING COSTS	$5,515	$4,300	$ 4,025
	TOTAL FOR FORTY YEARS	$220,000	$172,000	$161,000

FIGURE 34. Life Cycle Cost summary for lawn sprinkler system.

For special studies other formats may be used. Figure 35 is an example of a total cost study of various HVAC systems under consideration for a high school. This example, based on annual owning and operating costs, is the fifth lowest initial cost bidder being selected. It is in the organized approach to total costs that the VE concepts differ from the conventional approach used by the majority of design and construction firms.

Use of Weighted Constraints

After selection of alternates on a cost basis other elements, not readily assigned dollar values, must often be considered, e.g., aesthetics, durability, salability, etc. The worksheets shown in Figures 36 and 37 are designed to aid in evaluating and selecting alternates that take these factors into consideration. Here again, the team approach to selecting and weighing the constraints is vital. One discipline or individual cannot realistically evaluate the diverse parameters involved.

The forms are used in the following way. The best of the alternate ideas from a previous worksheet are listed, and the criteria which will have an important impact on final acceptance are written across the top. Parameters such as reliability, maintenance, standard part or process, weight, aesthetics, salability, interchangeability, competition, quantity, part of a family of similar items, time, and state of the art are listed where applicable.

Each parameter is given a numerical weight between, say, 1 and 4, depending on its order of importance. If savings potential is the most important consideration, that parameter is assigned an index of 4; if marketability is deemed next in importance, it is given a 3 (or if it is considered as important as the savings potential it, too, may be given a 4) and so on. These "weights" are noted along the top of the worksheet.

Each idea is next evaluated against each parameter and is given a number between 1 and 4: 1 if it is poor and 4 if it is excellent. These numbers are then multiplied by the "weights" (1 to 4) at the top, and each horizontal line is added to obtain a "score," written in the extreme right-hand column. The ideas receiving the highest scores are retained for development in the proposal phase which follows.

Thus we see on Figure 36 that low cost and quiet operation are the factors having the greatest impact on the choice of the on-off switching device. Yet nine proposals rank similarly for these two factors. The third factor reduces the choice to two proposals. It is not until the last parameter is evaluated that the final choice emerges. From this example one can see the importance of judgment in assigning parameters and their weights—clearly, a team effort has many advantages.

In Figure 37, the choice of the metal pan stair is established at the outset because of the importance of cost over the other constraints.

Proposal Phase

The proposal phase, sometimes called the program planning and reporting phase, is the final step in the VE job plan. During this phase three things must be accomplished.

1. The group must thoroughly review all alternate solutions being proposed to assure that the highest value and significant savings are really being offered.
2. A sound proposal must be made to management. The group must consider not only to whom it must propose, but also how to propose the solutions most effectively.
3. The group must present a plan for implementing the proposal. This action is critical, for if the proposal can't convince management to make the change, all the work goes for naught.

The proposal is now in the hands of management. One point bears further clarification at this time. The VE team can only recommend; it is up to management to concur and to the designers to make the final changes.

A complete report includes:

1. A brief description of the project studied
2. A brief summary of the problem
3. The results of your functional analyses, showing existing and proposed designs
4. Technical data supporting your selection of alternates
5. Cost analyses of the existing and proposed designs
6. All associated data, quotations, and suggestions
7. Sketches of before and after designs, showing proposed changes clearly (plans marked to show proposed changes are generally acceptable)
8. A description of the tests used to evaluate your proposed design, and how the idea passed the tests
9. Acknowledgment of the contributions by others
10. A summary statement listing all the reasons for accepting the proposal, and any actions required for implementation

Figure 38 is a suggested format for submitting proposals.

Appendix B contains examples of completed proposals. Example 1 was developed during a VE workshop on the exterior wall section of a typical small office building used throughout an agency. Due to time limitations, exact identification of life-cycle costs were

Bidder Type of Equipment	1 SCRT Electric	2a SCRT Electric	2b SCRT Oil & Electric	3a SCRT Electric	3b CESRT Oil & Electric	4 CESRT Oil & Electric
Base bid	$ 959,000	1,148,700	1,166,700	1,240,000	1,325,000	1,254,000
Interface adjustments	——	——	——	——	——	——
Electrical installation	156,000	146,000	130,000	146,000	58,000	36,000
Plumbing installation	——	——	——	——	10,000	10,000
Addition construction (walks and enclosures)	50,000	52,000	54,000	54,000	44,000	9,000
SUBTOTAL	$1,165,000	1,346,700	1,350,700	1,440,000	1,437,000	1,309,500
Deductive option, 1*	– 64,000	-66,171	-66,171	-17,000	-13,000	-39,310
SUBTOTAL	$1,101,000	1,280,529	1,284,259	1,423,000	1,424,000	1,270,190
Deductive option, 2†	-8,000	-28,731	-28,731	-58,000	-50,000	-40,000
TOTAL ESTIMATED FIRST COST	$1,093,000	1,251,798	1,255,798	1,365,000	1,374,000	1,229,790
Amortization (5.5%, 20 years) of estimated first costs	$ 91,500	104,800	105,100	114,300	115,000	102,900
Electricity	94,660	89,660	24,600	89,660	24,700	21,100
Oil	——	——	37,000	——	37,000	45,900
Water	——	——	——	——	360	720
Maintenance	21,000	25,200	28,270	27,300	18,400	17,700
TOTAL ESTIMATED ANNUAL OWNING AND OPERATING COSTS	$ 207,160	219,660	195,400	231,260	195,460	188,320

*Deductive option for 1-year guarantee instead of a 5-year guarantee.
†Deductive option for 1-year maintenance instead of 5-year maintenance.
Note: Annual utility costs are based on normal school operation and are estimated for water heating, air-conditioning, and heating for the entire school.

FIGURE 35. Cost estimate summary for HVAC systems in senior high school. SCRT, self-contained rooftop system; CESRT, central station rooftop system.

IDEAS	Silent (4)	Low Cost (3)	Adaptable to Existing Function, Box & Circuit (2)	Minimum Associated Components Required (1)	Score	Rank
Silicon rectified	4	3	3	2	32	
Knife switch	4	4	2	3	35	5
Transistor	4	3	3	2	32	
Capacitor	4	4	2	2	34	
Rheostat	4	3	4	4	36	4
Vacuum tube	4	4	1	1	31	
Photocell	4	4	2	2	34	
Thermistor	4	3	1	1	27	
Bi-metallic	4	2	1	1	23	
Diode	4	4	1	2	32	
Transformer	4	1	1	2	20	
Mag. amplifier	4	1	1	2	20	
Insulated thumbtack	4	4	3	3	37	3
Mousetrap	1	4	1	2	19	
Saltwater capsule	4	4	4	3	39	2
Wheatstone bridge	4	3	3	2	32	
Stapler	2	3	2	1	23	
Icepick	3	4	2	2	31	
Clothespin	4	4	2	2	34	
Hand on clock	4	1	3	2	24	
Magnet	4	4	4	4	40	1

FIGURE 36. Weighted constraints chart for switching system. 1, poor; 2, fair; 3, good; 4, excellent.

IDEAS \ CONSTRAINTS	Cost	Maintenance	Appearance	Erection Time	Stiffness	Flexibility	Score	Rank	
WEIGHT (1-10)	10	4	2	8	7	9			
1 WAY/DEEP	3.5	3	2	1	3	3			
	35	12	4	8	21	27	107	3	107
1 WAY/FLAT	3	3	2	3	2	3			
	30	12	4	24	14	27	111	2	111
WAFFLE	2	2	4	2	2	3			
	20	8	8	16	14	27	93	5	93
METAL PAN	4	3	1	4	3	4			
STAIR	40	12	2	32	21	36	143	1	143
CONCRETE STAIRS	3	3	2	1	3.5	2			
	30	12	4	8	25.5	18	96.5	4	96.5

FIGURE 37. Weighted constraints chart for structural system. 1, poor; 2, fair; 3, good; 4, excellent.

PROJECT _____   DATE _____

ITEM _____ PROJECT NO._____

Summary of Change (Description)

Before $ _____   After $ _____

ESTIMATED COST SUMMARY (See attached cost estimates)

	NO. OF UNITS	UNIT COST	TOTAL
A. Original	_____	_____	_____
B. Proposed	_____	_____	_____
C. Initial savings	_____	_____	_____
D. Life-cycle costs annual savings	_____	_____	_____
E. Life-cycle costs gross savings	_____	_____	_____
F. Total	_____	_____	_____

PERCENT SAVINGS INSTANT _____

PERCENT SAVINGS LIFE-CYCLE _____

FIGURE 38.   Form for submitting VE proposal.

not developed. Example 2 is a completed proposal developed during a recent study of a highway project in which drainage structure costs appeared excessive and were isolated as a high-cost area.

Problems and Roadblocks in Application

Some areas must be approached cautiously to avoid potential roadblocks in implementing a VE program. The principal ones are:

1. Design personnel must be assured that the VE effort will not alter their decision-making authority. The VE group has authority only *to recommend.* The VE effort should not be used to circumvent line authority.
2. Management must voice active support, or better still, actively participate in the effort. Without these actions, experience shows that some middle management people tend to withhold information from the VE staff and to pass over VE recommendations without an honest appraisal.
3. Management should periodically review the program to assure:
   a. That the effort is not being used for tasks outside assigned VE duties
   b. That the effort is being used not solely for publicity, but principally for the development of technically oriented proposals
   c. That the effort is not antagonizing the organization by simply "nit-picking" their work without significantly contributing to overall cost effectiveness

TECHNIQUES OF PROJECT SELECTION

Before the VE methodology can be applied, items of potentially low value and high cost must be isolated. This identification process is the "art" of the VE approach. There are some 3000 to 5000 cost items involved in a modern facility. To find those few items or areas that represent the bulk of unnecessary costs is not easy. This task can be facilitated by using a number of techniques such as breakdown analysis, cost models, functional analysis, analysis of previous study areas, and study of life-cycle cost impact.

Breakdown Analysis

In this analysis, the systems, subsystems, and special equipment are ranked from highest to lowest in terms of total cost per unit in order to portray distribution of expenditures. The analysis is further refined by breaking down unit costs into functional areas such as electrical, mechanical, structural. In addition, use is made of principles expressed by Pareto's law (see Figure 8).

For example, a VE study was recently undertaken on a highway project. In order to use an organized approach to project selection, the cost estimate was broken down into major cost elements, numbering approximately 100. Twenty of these elements, making up about 20% of the total, were selected for in-depth study. They represented 86% of the total costs. Figure 39 is the worksheet used for the project.

Cost Models

The use of models for unit cost analysis enables systems to be broken down to simplify selection of areas for additional study. Thus the potential for generating VE improvements can be studied at several levels, starting with unit material and labor costs and proceeding to design, construction, and other costs.

The general-purpose cost model form is shown in Figure 40, and the form showing square-foot costs of building elements is illustrated in Figure 41. Examples of the use of the latter cost model are found in Figure 42 (small office building), Figure 43 (high school), Figure 44 (laboratory building), Figure 45 (communications facility), Figure 46 (telephone exchange), and Figure 47 (living quarters).

1. Cleaning and grubbing	$	255,400
2. Mobilization		402,720
3. Excavation — Cut	Earthwork	3,128,170
4. Embankment — Fill		2,812,660
5. Select borrow		556,600
6. Select granular fill		372,260
7. Subbase course granular material		703,950
8. Chain-link fence		117,000
9. Trench and culvert excavation		123,760
10. Corrugated metal pipe		150,000
11. Class A concrete		726,000
12. Class B concrete		707,850
13. Metal reinforcement for concrete pavement		202,975
14. Reinforcing steel		286,452
15. Structural steel		615,636
16. Guard rails and anchorage		92,730
17. P.C.C. pavement		2,040,500
18. Topsoil		211,610
19. Base course asphalt concrete		158,000
20. Guide rails and other safety features		173,000
	TOTAL	$13,837,273

Approximately 20 out of 100 items have the
following cost distribution:

Total cost	100 items	$16,108,368
Cost of selected items	20 items	$13,837,273

Percentage of total cost for 20 items = 13.8/16.1 X 100 = 86%

FIGURE 39. Breakdown analysis of typical highway project.

Cost targets for each item in the model should be established to assist in determining VE potential. This method of analyzing the design, specifications, and other requirements can be used as a means for isolating cost-saving potential. The general-purpose cost model for initial building costs (Figure 41) is used to graphically portray these target costs. The basic cost models are based on team expertise and/or estimates of similar buildings conditioned by past VE studies and on cost experience of an organization doing over $1 billion in annual project estimates. The basic cost model is an idealized facility meeting all functional and aesthetic requirements at lowest reasonable costs.

When similar facilities are to be value engineered, the actual costs of the major elements are compared with the basic cost model.

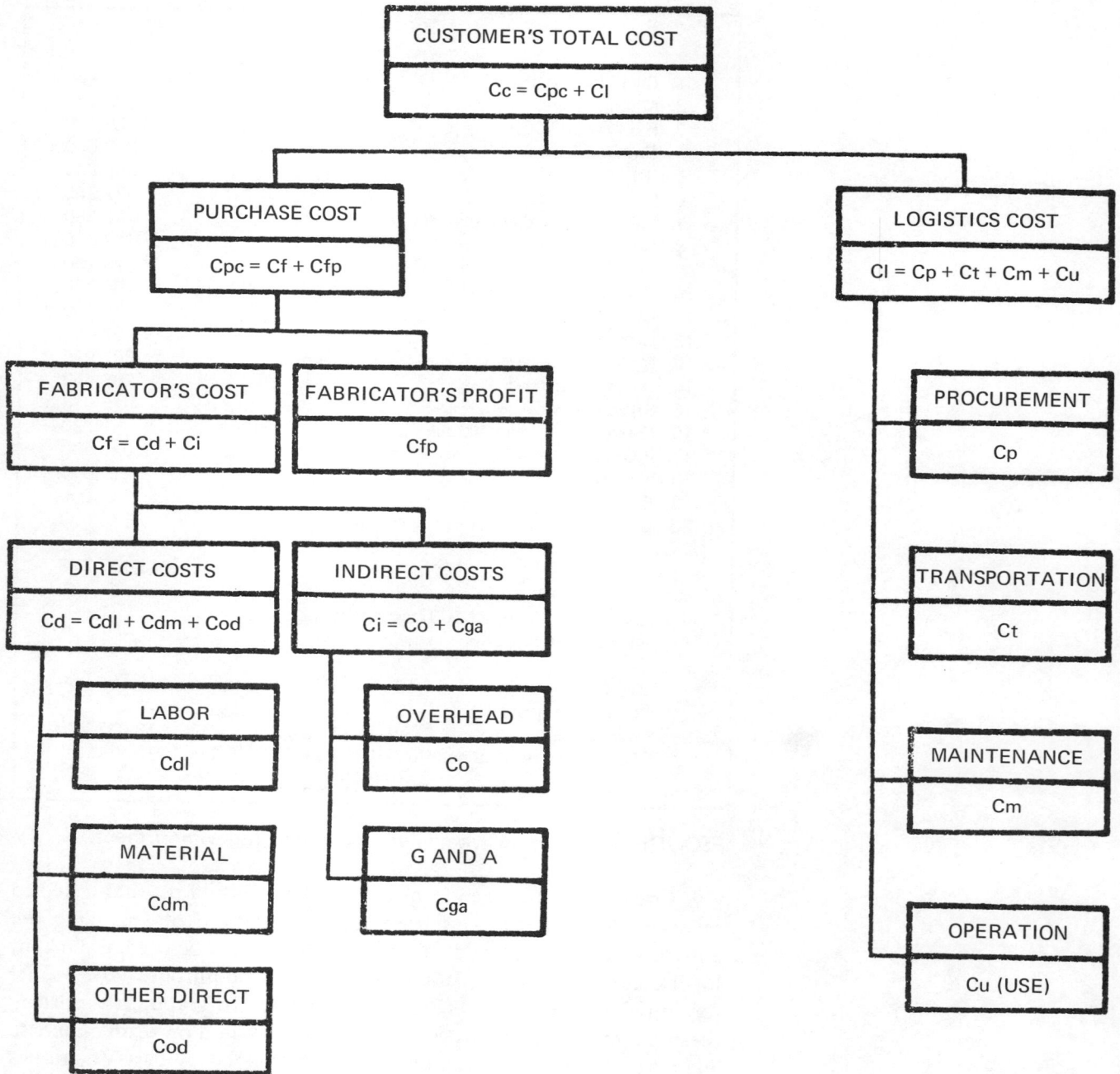

```
CUSTOMER'S TOTAL COST
Cc = Cpc + Cl
```

```
PURCHASE COST
Cpc = Cf + Cfp
```

```
LOGISTICS COST
Cl = Cp + Ct + Cm + Cu
```

```
FABRICATOR'S COST
Cf = Cd + Ci
```

```
FABRICATOR'S PROFIT
Cfp
```

```
DIRECT COSTS
Cd = Cdl + Cdm + Cod
```

```
INDIRECT COSTS
Ci = Co + Cga
```

```
PROCUREMENT
Cp
```

```
TRANSPORTATION
Ct
```

```
LABOR
Cdl
```

```
OVERHEAD
Co
```

```
MAINTENANCE
Cm
```

```
MATERIAL
Cdm
```

```
G AND A
Cga
```

```
OTHER DIRECT
Cod
```

```
OPERATION
Cu (USE)
```

FORMULA: $Cc = \{[(Cdl + Cdm + Cad) + (Co + Cga)] + Cfp\} + (Cp + Ct + Cm + Cu)$

FIGURE 40.   General-purpose cost model.

Adjustments are made in regard to quality levels. Where significant differences are found between actual costs and the basic cost model without special conditions involved to justify these differences, the areas are isolated for functional analysis.

For example, Figure 42 was used while conducting a VE study of a small office building used throughout an agency. The model was analyzed, and the architectural elements' basic cost of *$5.55* vs. their actual cost of *$9.36* definitely isolated this area of savings potential. The exterior wall section and roof were selected for this category for in-depth analysis. These elements were selected over interior partitions because of special conditions requiring more interior partitions for the actual building than the basic building. For complete study see Appendix B, Example 1.

Figure 43 is a model developed for cost control and VE guidance for a large school. This model was developed during the initial phases of design. It was used as a tool by the construction manager to have the designers design to costs rather than ending up costing a design. That is, periodic estimates were made on components and compared with the cost models. It proved a useful tool.

Figure 44 is a comparative cost model of a teaching hospital and a medical science building. It proved useful as a guide to isolate potential saving areas for the proposed medical facility. In this case, the team decided to compare the functional cost model of the medical science building and teaching hospital with extensive laboratory facilities recently value engineered by the firm. The team knew the cost of the teaching hospital represented "good" value. When comparing cost models, the team noted the structural system of the actual building exceeded that of the basic teaching hospital by a significant amount ($8.23 vs. $7.15). This, even though the basic building was designed to meet earthquake criteria. As a result, the team reviewed the structural area and decided the structural costs for the penthouse appeared excessive for the functions being performed and should be isolated for study. Other areas were found to have questionable cost ratios. The team looked at the redundant heating system as the reason for the difference in HVAC costs ($9.21 vs. $8.25). The team felt the typical lighting system for the lab area could be reduced to bring the lighting costs ($4.15 vs. $3.30) closer together. The last area was selected because there was a large difference in equipment costs ($4.69 vs. $2.50). The major item of equipment was the fume hoods. An in-depth study on the hoods was initiated.

The actual proposals for improving structural, HVAC, lighting, and fume hood costs are shown in Examples 14 to 17, below under "Examples of Actual Results." Approximately $200,000 was removed from the facility cost by the VE effort.

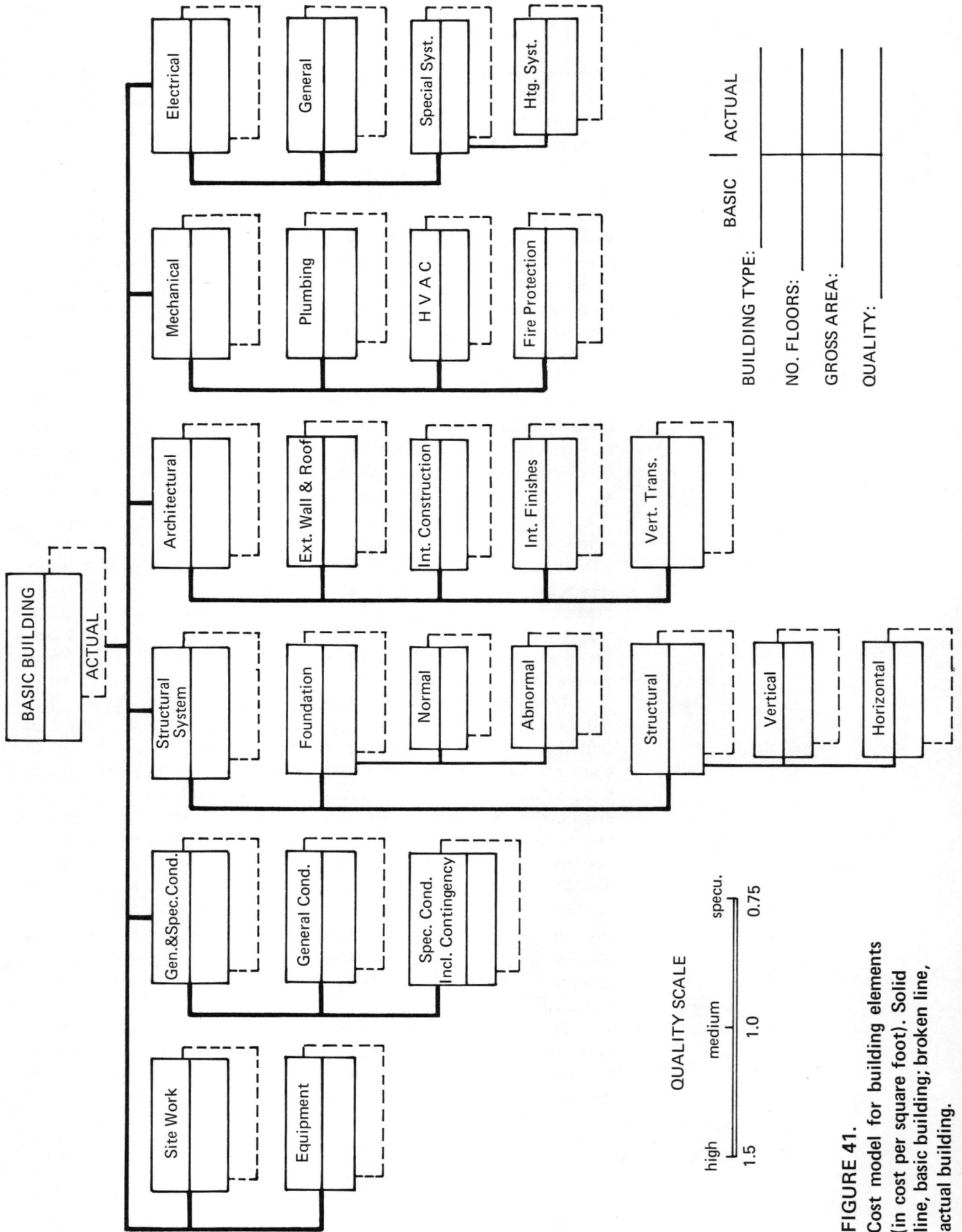

BASIC BUILDING | ACTUAL

Gen.&Spec.Cond.
- General Cond.
- Spec. Cond. Incl. Contingency

Site Work

Equipment

Structural System
- Foundation
  - Normal
  - Abnormal
- Structural
  - Vertical
  - Horizontal

Architectural
- Ext. Wall & Roof
- Int. Construction
- Int. Finishes
- Vert. Trans.

Mechanical
- Plumbing
- H V A C
- Fire Protection

Electrical
- General
- Special Syst.
- Htg. Syst.

BUILDING TYPE: _____
NO. FLOORS: _____
GROSS AREA: _____
QUALITY: _____

BASIC | ACTUAL

QUALITY SCALE

high — medium — specu.

1.5 — 1.0 — 0.75

FIGURE 41.
Cost model for building elements (in cost per square foot). Solid line, basic building; broken line, actual building.

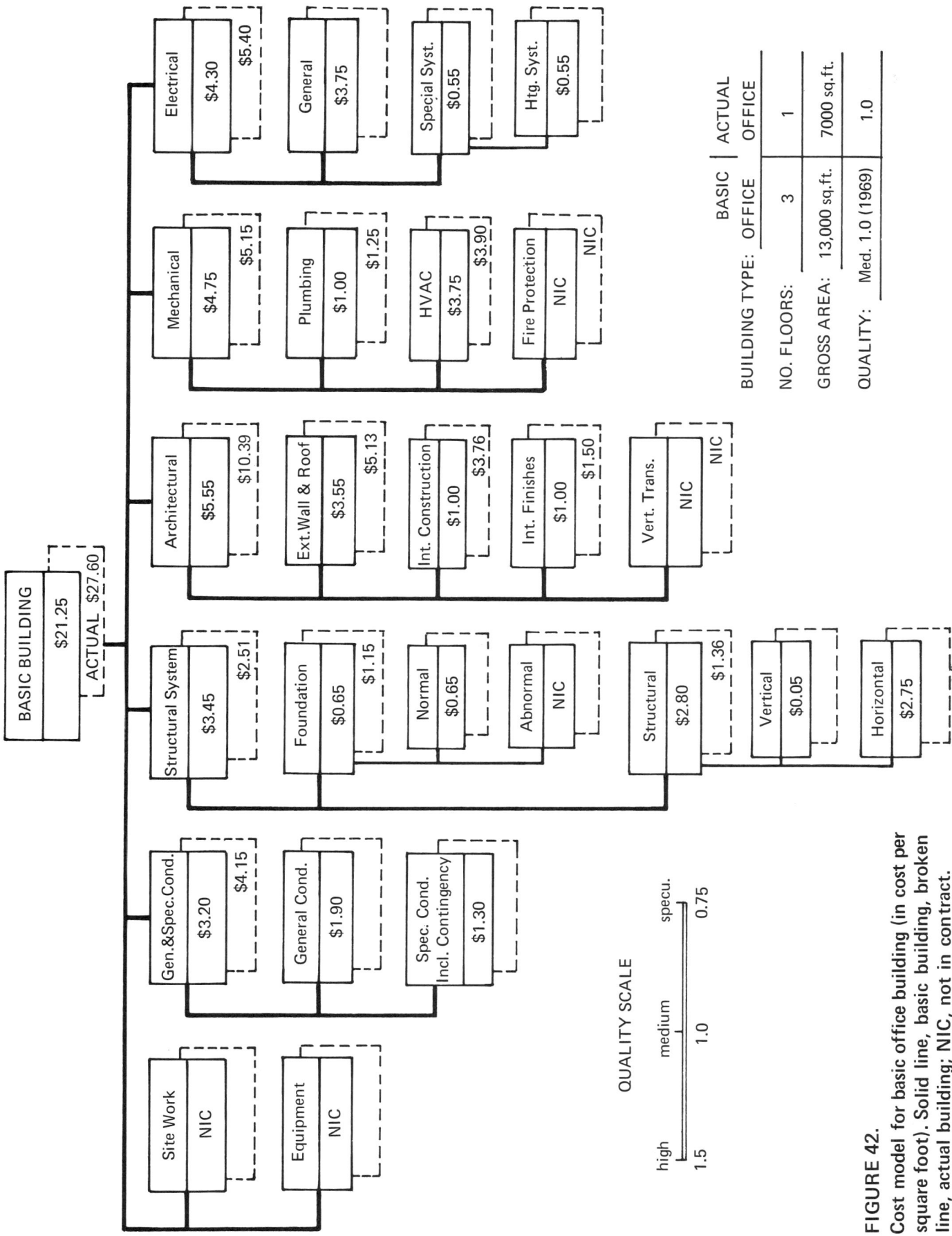

FIGURE 42.

Cost model for basic office building (in cost per square foot). Solid line, basic building, broken line, actual building; NIC, not in contract.

The following data appear within the figure:

	BASIC OFFICE	ACTUAL OFFICE
BUILDING TYPE:		
NO. FLOORS:	3	1
GROSS AREA:	13,000 sq.ft.	7000 sq.ft.
QUALITY:	Med. 1.0 (1969)	1.0

QUALITY SCALE

high — medium — specu.
1.5 — 1.0 — 0.75

BASIC BUILDING $21.25 — ACTUAL $27.60

Site Work NIC
Equipment NIC

Gen.&Spec.Cond. $3.20 $4.15
General Cond. $1.90
Spec. Cond. Incl. Contingency $1.30

Structural System $3.45 $2.51
Foundation $0.65 $1.15
Normal $0.65
Abnormal NIC
Structural $2.80 $1.36
Vertical $0.05
Horizontal $2.75

Architectural $5.55 $10.39
Ext.Wall & Roof $3.55 $5.13
Int. Construction $1.00 $3.76
Int. Finishes $1.00 $1.50
Vert. Trans. NIC NIC

Mechanical $4.75 $5.15
Plumbing $1.00 $1.25
HVAC $3.75 $3.90
Fire Protection NIC NIC

Electrical $4.30 $5.40
General $3.75
Special Syst. $0.55
Htg. Syst. $0.55

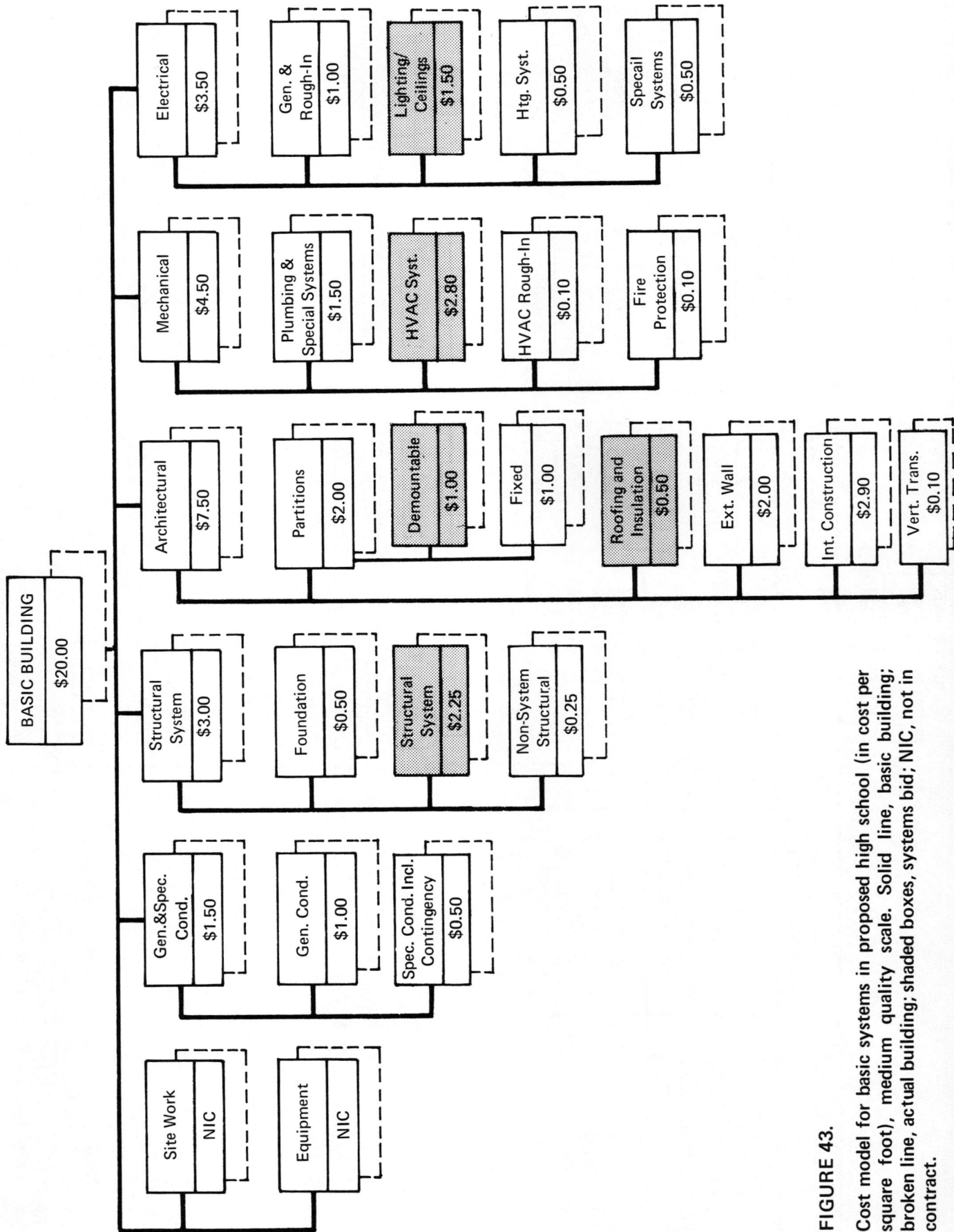

FIGURE 43.

Cost model for basic systems in proposed high school (in cost per square foot), medium quality scale. Solid line, basic building; broken line, actual building; shaded boxes, systems bid; NIC, not in contract.

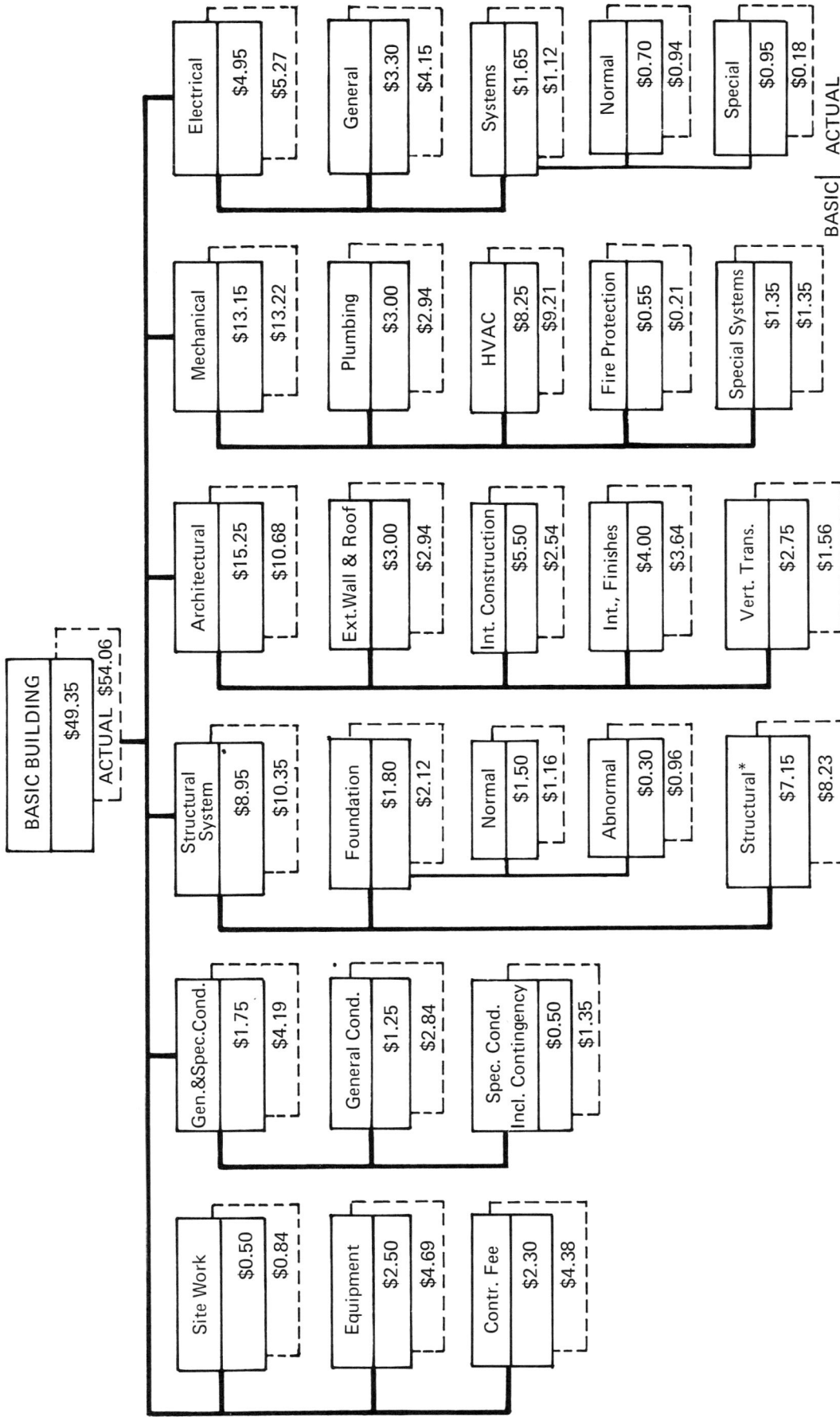

FIGURE 44.

Cost model for hospital and medical science building (in cost per square foot). Solid line, basic building; broken line, actual building; *, earthquake area.

Figure 45 was developed as a project selection guide for a recent study of a small repetitive ($300,000) communications building. The VE team brainstormed the functions required to be performed in the facility. An idealized cost per square foot was assigned each functional area. These represented optimum costs. For example, the group felt that the optimal structural costs for this facility should be $5.50/sq. ft. That is, if the actual building cost came to this amount, very little savings potential would exist (provided needed requirements were fulfilled). After finalizing the cost model, the records of the firm were reviewed for similar projects. Functional cost areas were reviewed to assure that the idealized costs developed were realistic.

As a result of the above exercise, the study effort was focused on the structural area, where the difference between the idealized and actual ($5.50 vs. $11.50) was the greatest. Approximately $50,000 in savings resulted from the study and future designs were changed to implement these savings.

Figure 46 is the cost model used as a project selection guide in a recent VE study on a telephone exchange facility of approximately 60,000 sq. ft. Basic building costs were derived from a recently completed facility whose design had been concurrently value engineered by a knowledgeable VE team. As can be noted from the chart, the two areas of greatest savings potential are the mechanical ($5.63 vs. $10.32) and architectural ($4.62 vs. $8.00). In addition, the water distribution system was reviewed ($2.00 vs. $2.50) and the lawn sprinkler system isolated for study. For results of the VE analysis see Examples 18 to 23, below, under "Examples of Actual Results."

Figure 47 is the cost model used for a VE analysis of living quarters. The review team developed an idealized model of the cost of the most economical living quarters. From the cost model comparison, efforts were focused on reducing the cost impact of the structural and architectural areas of the design.

To enable comparison of facilities in different areas and with differing aesthetic requirements, a quality scale is included in some of the cost models. This scale enables the VE team to adjust costs of buildings with different quality levels. For example, a commercial office building in a surburban area would have a quality level of medium (1.0). An office building for a large corporation in a metropolitan location would have a quality level of high (1.5). As much as a 50% difference in costs would have to be considered when comparing costs of these two facilities.   (See Figure 42).

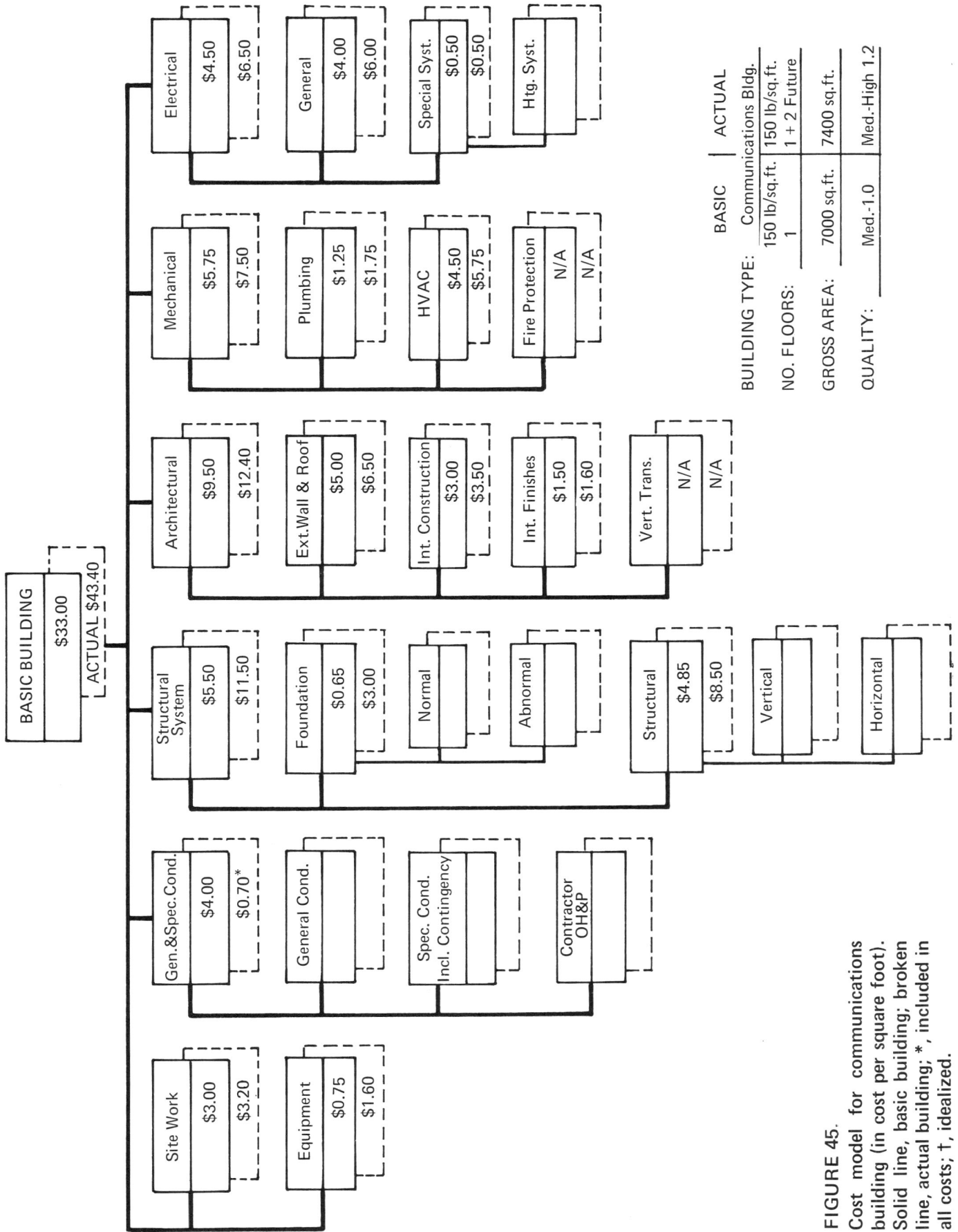

	BASIC	ACTUAL
		Communications Bldg.
BUILDING TYPE:	150 lb/sq.ft.	150 lb/sq.ft.
NO. FLOORS:	1	1 + 2 Future
GROSS AREA:	7000 sq.ft.	7400 sq.ft.
QUALITY:	Med.-1.0	Med.-High 1.2

FIGURE 45.
Cost model for communications building (in cost per square foot). Solid line, basic building; broken line, actual building; *, included in all costs; †, idealized.

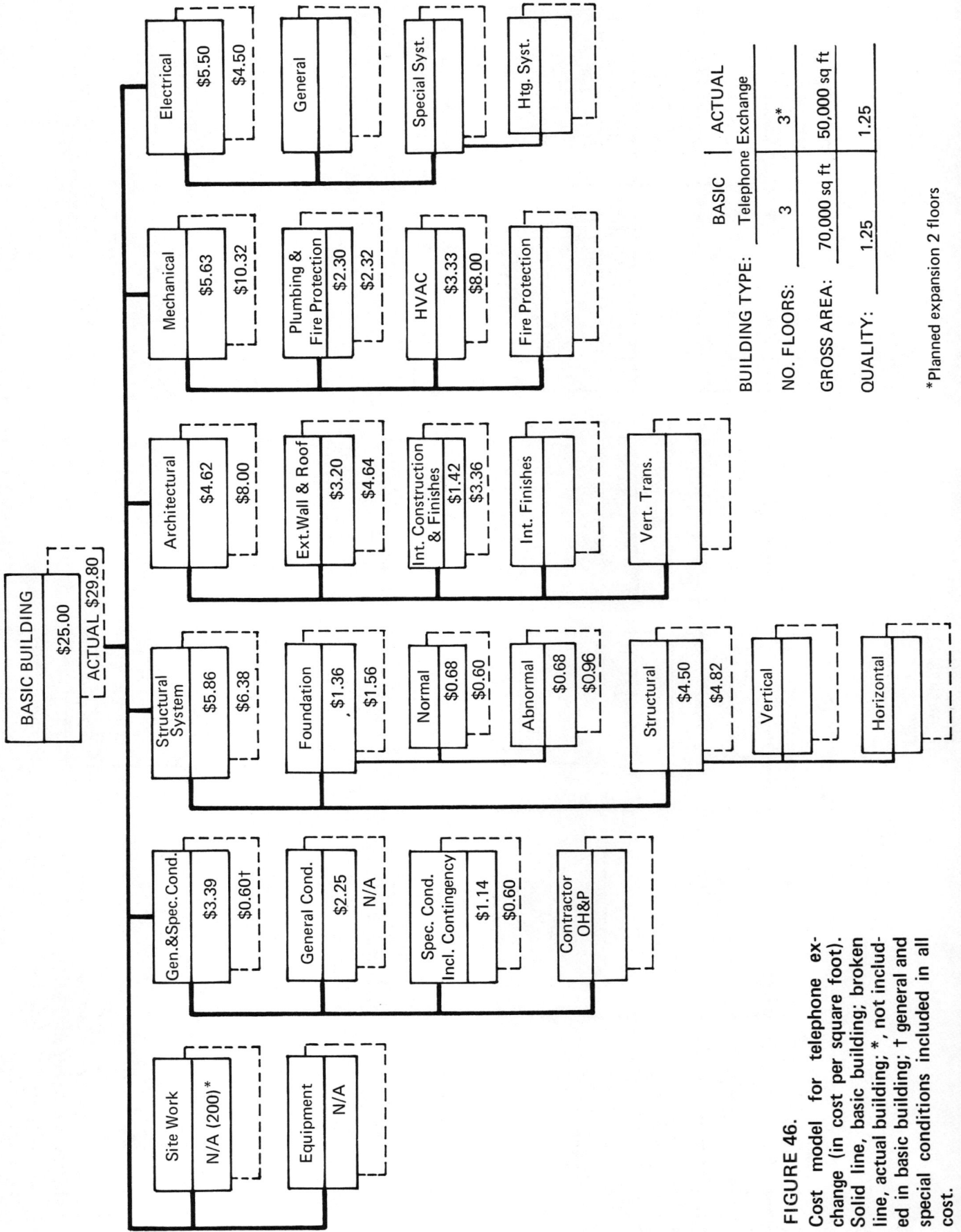

FIGURE 46.

Cost model for telephone exchange (in cost per square foot). Solid line, basic building; broken line, actual building; *, not included in basic building; † general and special conditions included in all cost.

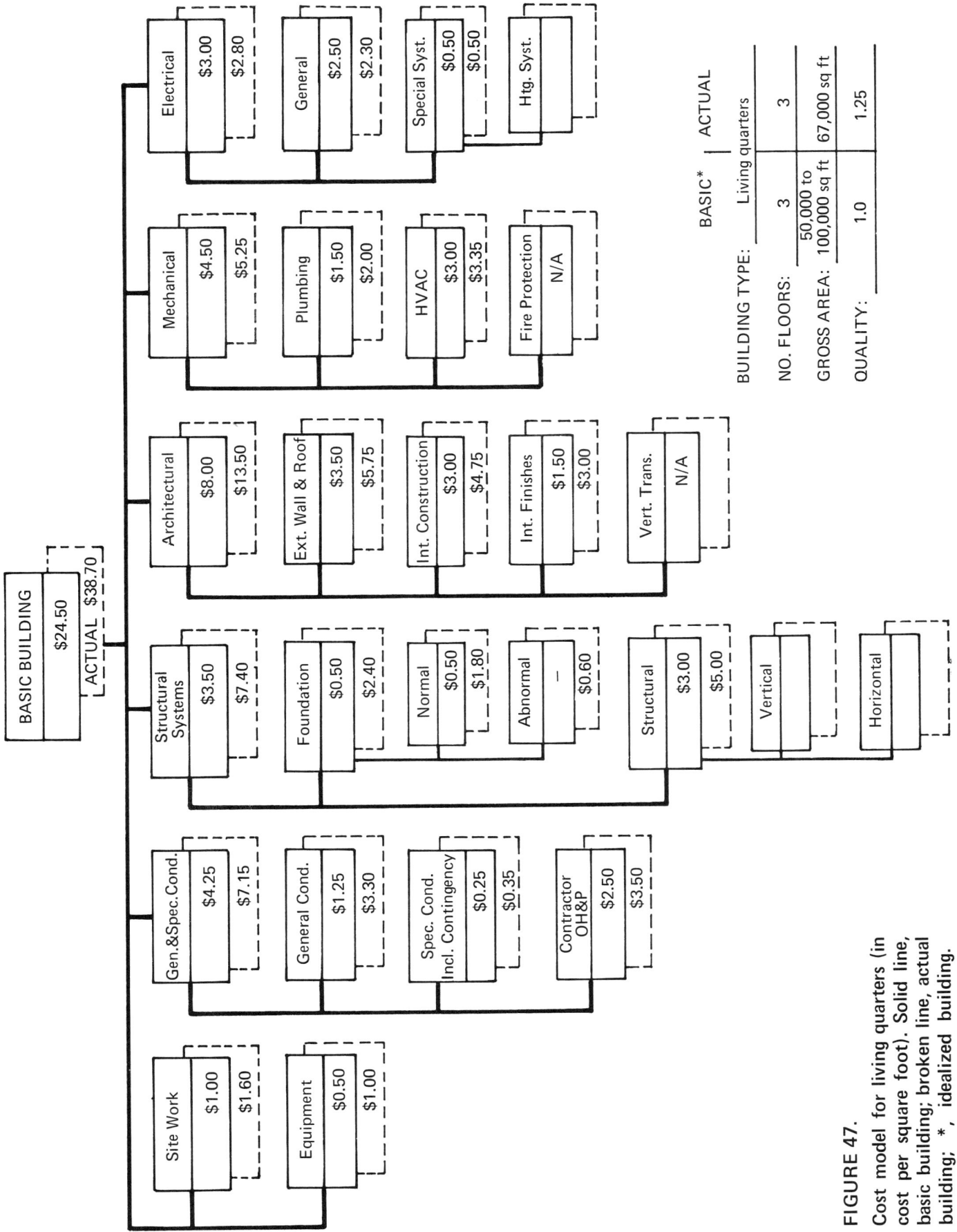

FIGURE 47.

Cost model for living quarters (in cost per square foot). Solid line, basic building; broken line, actual building; *, idealized building.

Functional Analysis

The functional analysis technique, when used to identify unnecessary or redundant functions, quickly isolates areas with good potential for VE studies (see "VE Job Plan," above). By adding the cost of performing the functions, and comparing these with the estimated worth of performing the basic function, areas of inadequate value can be identified (see "Information Phase"). The function-cost-worth approach, when combined with breakdown analysis and use of cost models, is the most important tool available for selecting meaningful projects for in-depth analysis. The worth of performing the functions is derived from several sources. One source might be from cost targets established from experience. Another could be information from similar designs. For example, a home can be air-conditioned and heated for from $2 to $3/sq. ft. This figure can be used as a cost target for HVAC in offices and barracks. Then, if an office building is analyzed, and costs of these functions are estimated at $7.50, this area should be isolated as a VE project. The reasons are that the amount of refinement required from a home to an office building should not be ($7.50/$2.50 = 3) "worth" three times.

Note: The functional analysis of the exterior wall section isolated by the cost model is included in Appendix B, Example 1. From this it is apparent the cost to achieve the function "control elements" should be improved because of a high cost/worth ratio of over 6.

Previous Study Areas

After the VE program is in operation, the areas previously studied provide an excellent source of information for selecting projects. Analysis of data from organizations with formal VE programs and the experience of the author suggest that approximately 30% of savings developed by VE analysis can be traced back to results from previous studies.

Life-Cycle Cost Impact

The emphasis on first cost and the resulting failure to include the total effect of related cost elements is probably the greatest shortcoming in today's planning, programming, and designing of facilities. These hidden costs have a considerable impact on cost of ownership. In addition to first cost, life-cycle cost impact includes:

1. Maintenance and operation costs
2. Money charges, such as interest, insurance, etc.

3. Future income or needs, such as rentable or usable space, future expansion
4. Fringe costs not subject to ready dollar analysis, such as aesthetics and durability
5. Ownership logistic costs such as material delivery, serviceability, personnel access, etc.
6. Service and operating personnel costs, such as janitorial services, operating personnel (i.e., engineers, doctors in hospitals), etc.
7. Real estate and property taxes.

COSTS AND FINANCES

Project Cost

The costs of the proposed union hall, with its union offices, multi-purpose hall, classroom facilities, and rental office space, have been estimated below.

The total project cost of $1,467,500 is the estimate to get the building fully constructed and ready for occupancy. This figure is exclusive of land costs, since both the main parcel on which the building will sit and the auxiliary parking parcel across the street are currently in full union ownership, and thus do not represent a cost of the presently conceived project.

Land cost	------
Building cost	$1,200,000
Parking (34,800 sq ft @ $1.50/sq ft)	52,200
Landscape allowance	10,000
Architect's fee (6%)	75,800
Developer's fee (1%)	12,600
Legal fees (2%)	25,200
Total land and building cost	$1,375,800
Interest during construction	68,800
Loan placement charge	6,900
Taxes during construction	16,000
TOTAL PROJECT COST	$1,467,500

FIGURE 48. Form for project budget.

Figure 48 is a typical example of the format used in developing a project budget. It illustrates a failure to take into account all the costs that will accrue to the owner over the life cycle of his project, especially those for maintenance and operations.

Miscellaneous Analyses

1. High support cost. Material and equipment requiring high maintenance, operation, or replacement costs are likely candidates for VE study.
2. Long-lead-time items. Low-value, long-lead-time items are good areas to consider for study. VE items frequently offer improved availability or delivery.
3. Repetitive items. Another area for savings is where a number of similar items appears in the design. Even a small savings applied to numerous items can be worthwhile. These savings are especially worthwhile when dealing with repetitive designs.
4. State of the art. Designs that are pushing the state of the art or are of recent origin without much history, offer a potential for VE. Many times their cost far outweighs the functions being performed. In addition, problems can occur with supply and replacement.
5. Complexity of the item. Complex items frequently indicate opportunity for improved value.
6. Staid or unchanged designs. Items which have remained unchanged for several years, and lack improvements made possible through technological advances, offer significant saving potential.

Figure 49 illustrates a building cost model which covers the major cost areas for the life cycle of an average building. Unlike the general-purpose model (Figure 40) that is useful in studying procurement of components, this model can isolate high costs among the total costs of construction and ownership. To isolate costs further in any one particular area, it may be necessary to break the general cost model into additional details using similar cost models for the subsystems. See previous discussion on the analytical phase of the job plan for examples of life-cycle cost impact.

It is interesting to note that commercial developers use the life-cycle approach to determine the return on investment. Future investments are selected from these determinations. Tables 1 through 5 are examples of the investor's approach to life-cycle costs. The VE team should strive to include individuals who can contribute this type of input. (See Appendix I).

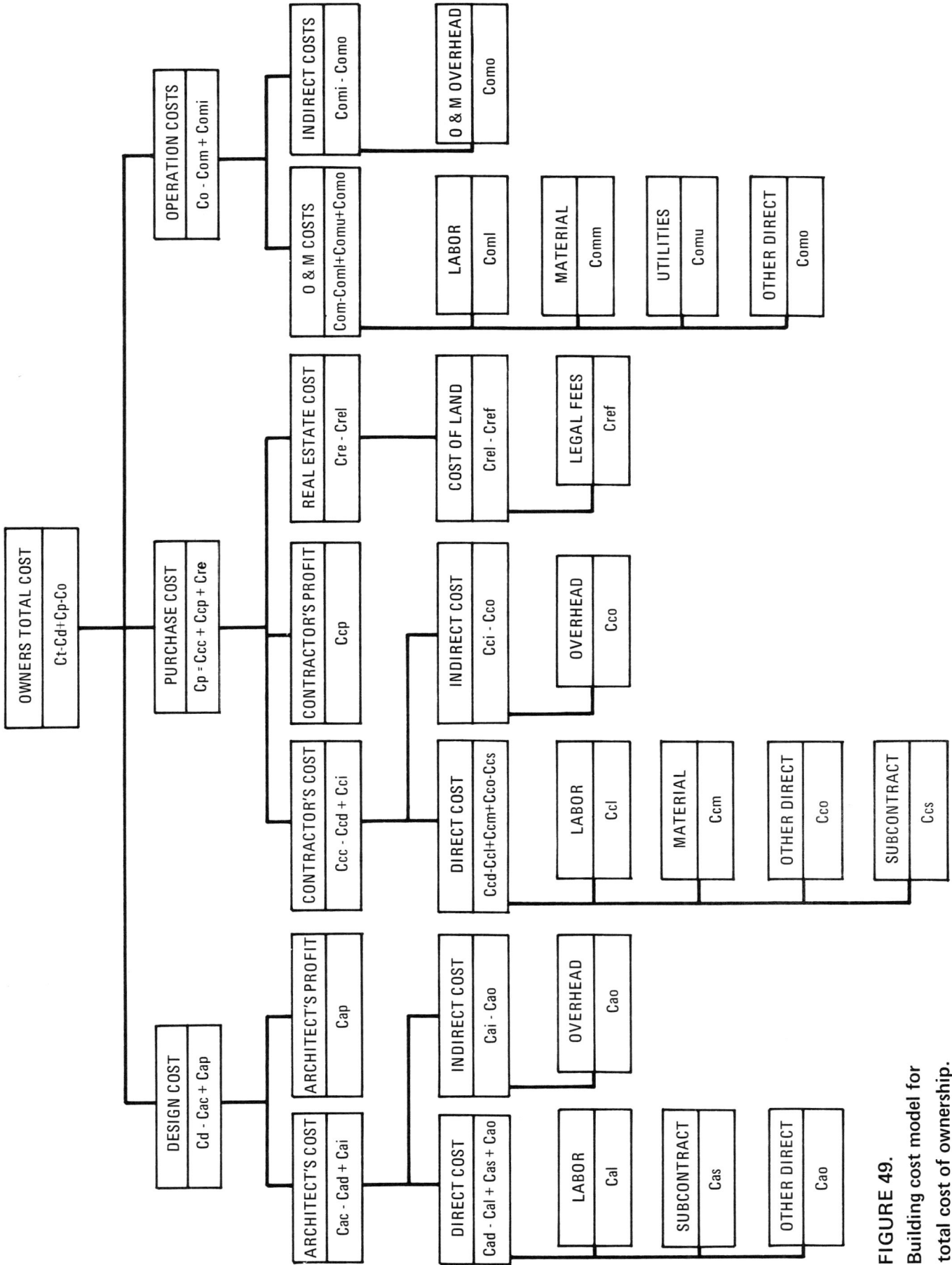

FIGURE 49.
Building cost model for total cost of ownership.

OWNERS TOTAL COST	Ct-Cd+Cp-Co	

DESIGN COST	Cd - Cac + Cap

ARCHITECT'S PROFIT	Cap

ARCHITECT'S COST	Cac - Cad + Cai

INDIRECT COST	Cai - Cao

OVERHEAD	Cao

DIRECT COST	Cad - Cal + Cas + Cao

LABOR	Cal

SUBCONTRACT	Cas

OTHER DIRECT	Cao

PURCHASE COST	Cp = Ccc + Ccp + Cre

CONTRACTOR'S COST	Ccc - Ccd + Cci

CONTRACTOR'S PROFIT	Ccp

REAL ESTATE COST	Cre - Crel

INDIRECT COST	Cci - Cco

OVERHEAD	Cco

DIRECT COST	Ccd-Ccl+Ccm+Cco-Ccs

LABOR	Ccl

MATERIAL	Ccm

OTHER DIRECT	Cco

SUBCONTRACT	Ccs

COST OF LAND	Crel - Cref

LEGAL FEES	Cref

OPERATION COSTS	Co - Com + Comi

INDIRECT COSTS	Comi - Como

O & M OVERHEAD	Como

O & M COSTS	Com-Coml+Comu+Como

LABOR	Coml

MATERIAL	Comm

UTILITIES	Comu

OTHER DIRECT	Como

Evaluation Criteria

While many opportunities for savings can be developed, normally it is not possible to conduct studies in each area. The items selected for evaluation should be determined on the basis of:

1. Probability of significant savings
2. Availability of time and resources
3. Probability of developing alternates of lower life-cycle costs
4. Probability of implementation

The following guidance is offered in the selection of projects to avoid nonproductive efforts.

Human factors. Many times the items or areas isolated for possible study were either placed in the original design by a principal design engineer or resulted from a stated requirement of the owner. To obtain a decision for VE study of an item under these conditions may be difficult because people don't resist change per se, they resist being changed.

Overlapping authority. In many areas of significant savings there is often more than one design discipline or organization unit involved. To get authority for a study involving several disciplines or organizational units may be difficult. In many instances, no effective communication between the disciplines or units has been established.

Project requirements. Many times the information about project requirements, such as who originated or why a particular item or criterion is included in the design is not available or is difficult to obtain. To select such an area for study, not knowing the reasons for inclusions, is generally not good practice since valid information about users' requirements is especially difficult to obtain.

Life-cycle costs. A number of unit costs must be estimated to develop life-cycle costs. Many times it is difficult to estimate costs for maintenance, operations, and replacement because few cost data exist. To spend the time to develop such data may not be productive.

APPLICATION OF VE JOB PLAN TO CONSTRUCTION

Limitations

To date, practically all VE has been applied either after high bids were received on a completely designed project or when a final estimate of completed drawings revealed cost problems. As noted in Figure 50, this is not the ideal time to apply VE because of lost time and increased costs to make changes. The time spent applying the VE methodology is added to the time required to complete the project. By applying the VE during the design phase, little or no additional time is required, and little cost is involved to implement changes.

As this figure shows, the savings potential differs between major design decisions and material and methods selections. If savings are to be realized on major design decisions, it is imperative to conduct VE studies before completion of working drawings. In fact, the optimum time to conduct a value review is after the preliminary submittal stage and before working drawings are started. A VE review of materials and methods selection has savings potential almost until completion of construction.

The sometimes violent resistance to changes by design personnel is a real limitation to the successful application of VE. For example, experience has shown that when a VE study is to be performed, designers who were involved in the original design tend to defer decisions on valid suggestions, to make quick design changes, or to withhold information required for objective appraisals. Also, once the VE study has started and study areas have been isolated, changes are sometimes predated in an effort to prevent the VE study from being associated with any positive results.

The principal reasons for resistance to changes recommended by VE studies are the lack of understanding of VE and, as noted earlier, a general resistance to any changes. The misconceptions are that VE is nothing more than the normal everyday work of a good engineer or the routine cost reduction performed by industry. On the contrary, VE is designed to be an additional technique for economy and for promoting greater cost effectiveness.

For example, there is a general lack of knowledge of the principles and application of VE in the construction area. These misconceptions are:

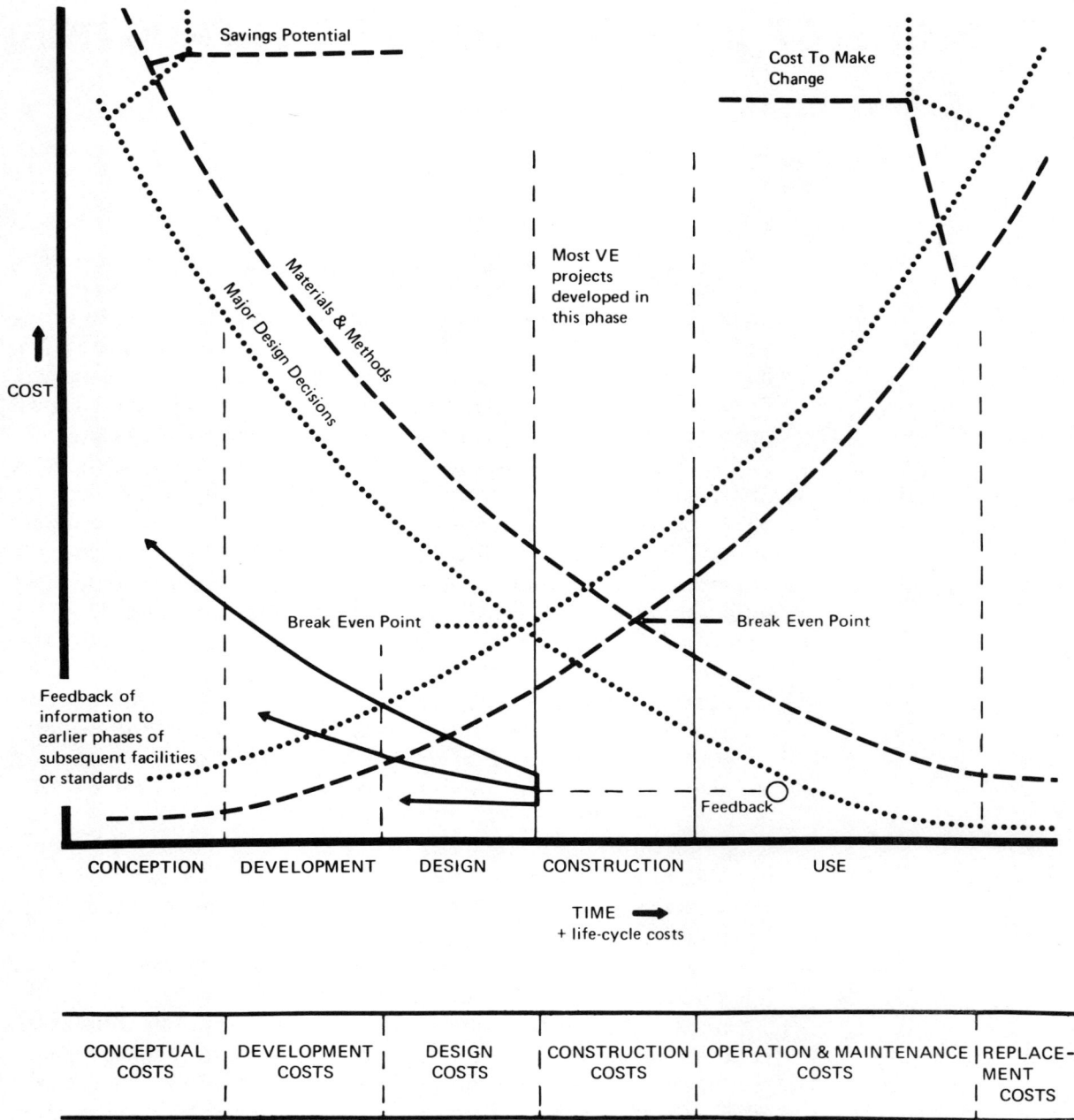

Savings Potential

Cost To Make Change

COST

Materials & Methods

Major Design Decisions

Most VE projects developed in this phase

Break Even Point

Break Even Point

Feedback of information to earlier phases of subsequent facilities or standards

Feedback

| CONCEPTION | DEVELOPMENT | DESIGN | CONSTRUCTION | USE |

TIME ➡
+ life-cycle costs

| CONCEPTUAL COSTS | DEVELOPMENT COSTS | DESIGN COSTS | CONSTRUCTION COSTS | OPERATION & MAINTENANCE COSTS | REPLACE-MENT COSTS |

FIGURE 50. Relation between savings potential and time in facilities construction.

Misconception	*Reality*
VE strives only to make things cheaper and does not consider quality.	Experience has proved that an organized creative approach by VE professionals has reduced costs without adverse effects on quality. In fact, quality often is improved.
VE is what the original designer does instinctively.	For the most part, today's designers are not trained in creative problem solving and follow conventional techniques that tend to isolate them from other disciplines. Training in VE can provide positive augmentation of this traditional approach.
A VE study is an unprofessional effort, as it tends to cast doubt on the integrity and competence of the original designer.	In practice, the VE effort seeks out additional design-oriented cost improvement. To be a true VE effort, it must assist designers on a professional level. Since the original designers retain the decision-making authority, no unprofessionalism enters into the system.
VE tends to delay the project or jeopardize the whole program.	If unnecessary costs exist, they should be revealed. If delays jeopardize a project because of VE changes those proposals should be dropped. However, experience has shown that VE changes favorably affect both cost and time. In addition, proposals dropped are available for future implementation on similar projects. If no VE efforts are attempted, no real augmentation of cost effective improvements would occur on a planned basis.

VE paperwork ends up costing more than the results.

Department of Defense records charge the cost of the VE program against the benefits derived. The immense saving from the defense department program (see page 11) does not support this criticism. In fact, the VE program is one, if not the only, program which ends up saving, not spending, money.

Ethics

Because of the critical nature of VE, there is difficulty convincing existing design professionals of its merit. In fact, the professional societies originally were opposed to the application of the VE concept to the design field. However, the legal counsels for the National Society of Professional Engineers and the American Institute of Architects reviewed the ethics rules and determined that under certain conditions a VE study of an existing design is ethical. These conditions are:

1. The original design agent's work has been terminated.
2. If the design is in progress, the design agent is informed of such study.
3. The study is conducted on a professional basis, preferably by professional people on a par with the design agent.
4. The original design agent receives extra compensation for any changes made as a result of the study.

See Appendix C for an article on the ethics of a VE review.

Modification of Standard Job Plan

To overcome the limitations of VE in construction, some modifications were made in the conventional VE approach.

Initially, it was recognized that the preferred place to do VE was during the design phase. However, industry response (architect-engineer) prevented application in a manner which would produce worthwhile results. Consequently, the contractors were solicited to support a VE incentive clause in construction contracts. They agreed and VE incentive provisions were placed in defense department construction contracts. Use of these provisions by construction contractors showed favorable results and awakened concern in the design profession. Slowly, some design organizations began to use VE methods.

For their first studies, VE study groups gathered the project plans, specifications, and design data sheets and went quickly through the job plan before submitting final proposals to the owner. This approach was not very productive. As was stated earlier, the original designers did not cooperate in revealing information and resisted implementation of proposals. In addition, significant changes were sometimes made in the requirements by the original decision makers during the study that negated much of the study findings. There were also cases where, in spite of VE efforts, the budget was exceeded and the project canceled.

The approach was then modified to initially conduct a briefing of involved personnel, outlining the scope and objective of the study. Next, all the available information, including cost data, was gathered. Finally, a VE study team was assembled. The team checked the cost estimate, and then conducted the information and speculative phases of the job plan. The objective was to verify costs and to develop a list of ideas of probable savings without expending too much time and money. Figure 51 is an idea listing sheet from an actual VE study. The idea listing and cost estimate were then reviewed with responsible owner-designated personnel to gain their input. The owner's representatives were given the authority to decide which ideas they wanted to implement, which would be developed by the VE team, and which would be discarded. It was interesting to note that owner interest was especially keen whenever the cost estimate revealed possible overruns in the budget. Appendix D is an idea listing developed for a completed design of a senior high school. The list contains the review comments and notes those items accepted and those selected for future in-depth studies. Subsequently, proposals were developed in draft form, and they were reviewed with those responsible for initial design decisions. Based on their comments, revisions were made to improve proposals or enhance the possibility of their implementation. This revised approach resulted in greatly improved implementation of proposals.

The basic philosophy behind the revised approach was to work, with owner cognizance, as closely as possible with the original decision makers on reducing costs. As a result, these persons became part of the change and shared credit for any savings. The final objectives were to keep the project within its scope and have good value.

The following is an outline of the revised job plan approach for the construction area:

I. Information phase
    A. Items required

DESCRIPTION	ESTIMATED POTENTIAL SAVING		DEVEL- OPMENT	RECOMMEND- ED PRIORITY		REMARKS
	INITIAL	LIFE CYCLE	MAN-HR	1ST	2ND	
A-3 Precast Stone Investigate the design and size with the possibility of reducing corners and greater use of lighter sections to facilitate placement. Reduce depth. Mechanical review team did not consider effect on A/C. In addition, investigate possibility of using precast sections to be a load carrying element. Savings of $4 sq ft appear possible.	$150,000	Unknown	88	88		
A-4 Suggest change of mechanical room to basement (see M-10). Change would affect following items (no changes considered due to weight reduction):						Tied to mechanical. Requires change in criteria issued to the A&E.
a. Precast screen: eliminate screen due to mechanical changes (see M-10)	75,000	Unknown	40	40		
b. Eliminate requirement for 150-lb load on roof and eliminate need for isolation pad design and mechanical room	50,000 ‾‾‾‾‾‾ 125,000	N/A				
c. Less cost for mechanical room in basement	-65,000 ‾‾‾‾‾‾ 60,000 (Net structural)	N/A $180,000				

FIGURE 51. Idea listing for office building. For detailed cost study of the relocation of the machine room, A-4, above, see Example 7, page 86.

1. Owner requirements given to architect and engineer
2. Plans and specifications
3. Cost estimates
4. Design data
5. Applicable codes, standards, and criteria
6. Applicable maintenance and operation data

B. Steps
1. Validation or preparation of cost estimate to assure compatibility of project scope and budget
2. Familiarization and breakdown of data into workable areas and preparation of cost models where applicable
3. Functional evaluation of areas to isolate high-cost, poor-value areas.

II. Speculative phase
A. Steps
1. Team review of project and generation of a comprehensive idea listing; suggestions for savings included on a tentative basis for review and input of owner
2. Brainstorming and review of idea listing and cost estimate by owner, original design personnel, and study teams
3. Development of final idea listing

III. Analytical phase
A. Steps
1. Evaluation of idea listing from speculative phase and selection of high-cost areas for in-depth study, owner-designer implementation, or deletion
2. Team brainstorming for generation of possible alternates for in-depth analysis
3. Investigation of alternates
4. Team review of analyses
5. Preparation of proposal drafts

IV. Proposal phase
A. Steps
1. Review of results of analytical phase with client to gain additional pertinent information and enhance implementation
2. Review and revision of proposals
3. Preparation of final proposals
4. Submission of report, including cost estimate review, to owner

An example of a typical proposal for a VE study is shown in Appendix E. Aside from the format, the reader can see how the scope and the evaluation of solutions are developed, and how the presentation of results is made.

EXAMPLES OF ACTUAL RESULTS

The following examples, numbered 1 through 24, are taken from studies that date back to early 1963 when the formal application of VE methodology was first implemented.

With each example is a brief narrative of its background, and in many cases, "before and after" diagrams. Examples 1 and 2 include the functional analysis worksheets developed during the application of the job plan. Many of the other examples were developed from expanded idea listings and/or VE training workshops where the client implemented ideas from the expanded idea listing and did not require further efforts. See "Modification of Standard Job Plan," above.

Additional examples of results of VE studies are presented in Appendix B.

▶ **Example 1: Deck for Refueling Pier**

VE Proposal

The VE group performed a study on the pier deck design. A number of galvanized steel corrugated sheets for slab forms have recently been marketed. Mechanical means are incorporated to allow the steel sheet form to act as a structural member. The result is the reduction of the required slab thickness for equivalent strengths. An alternate solution using this material with a protective coating (a figure of $0.30/sq. ft. for the coating was used) properly shored during construction ($0.50/sq. ft.) was developed and cleared in principle with the structural designers.

Recommended Protective Coating—Cleaning and Pretreatment in accordance with TT-C-490 Type I and finish coating (3/32") in accordance with 34Yd using coal tar primer and enamel MIL-C-15147C.

		Unit Cost			
Estimated Cost Summary		Material	Labor	Units	Total
A.	Original	$ 7,720	$ 9,830		$ 17,550
B.	Proposed		Lump sum		11,796
C.	Gross savings				5,754
D.	Implementing costs				500
E.	Net savings	$	$		$ 5,254
F.	Percent savings 30 %				

FUNCTIONAL ANALYSIS WORKSHEET

PROJECT __Deck for refueling pier__   ITEM __Pier deck design__

__Load design 100lb/sq ft, 1^k concentration load - 20ft span__   __Transmit load__

QTY.	UNIT	COMPONENT	FUNCTION			EXPLANATION	WORTH	ORIGINAL COST
			VERB	NOUN	KIND			
3555	sq ft	Timber section	Transmit	Load	B	Tensile load	$ 2,500*	$ 8,750
			Support	Concrete	S		-----	
		Shear plates	Transmit	Shear	S		-----	1,000
		Nails	Hold	Parts	S		-----	1,400
						Total tensile	2,500	$11,150
						Cost/value = 4.5		
3555	sq ft	Concrete section	Transmit	Load	B	Compressive load	$ 2,500*	$ 3,600
			Provide	Drainage	S			
			Provide	Safety	S	Curbing and fire		
		Rebar	Distribute	Stress	S	Prevent cracking		2,560
						Total compressive	2,500	6,160
						Cost/value = 2.5		
		Bolts	Hold	Parts	S			180
		Copper pipe	Remove	Water	S			60
						TOTAL =	$ 5,000	$17,550

*Based on a simple span (20-ft) wooden deck.

The area was isolated for study when a team member noted that the existing design was first introduced around 1938. Some better, less expensive way must have been developed since that time! The team performed a functional analysis (see worksheet above) and decided to focus attention on the wooden portion of the design since its cost/value ratio of 4.5 appeared excessive and indicated a significant savings potential. The team isolated some five alternate solutions which appeared promising, but decided to recommend only one alternate.

BEFORE

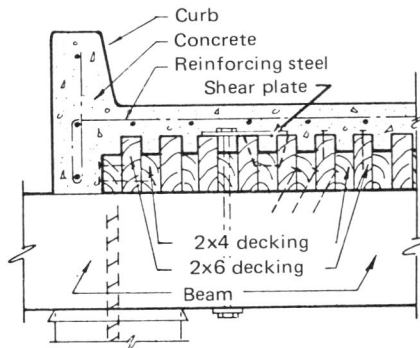

- Curb
- Concrete
- Reinforcing steel
- Shear plate
- 2x4 decking
- 2x6 decking
- Beam

AFTER

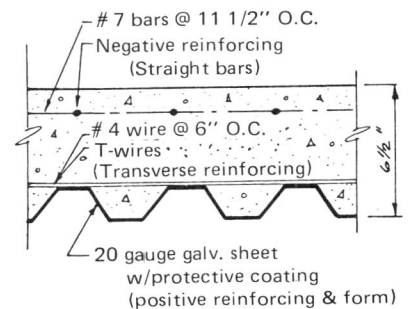

- # 7 bars @ 11 1/2" O.C.
- Negative reinforcing (Straight bars)
- # 4 wire @ 6" O.C.
- T-wires (Transverse reinforcing)
- 20 gauge galv. sheet w/protective coating (positive reinforcing & form)

77

▶ **Example 2: Roof Design for a Photographic Laboratory**

Description of Study

This area was isolated for study because of the redundancy of having two insulation elements performing the same function and because of the extra building height required for the rock wool batt insulation.

By developing an alternate which combines functions and eliminates unnecessary building height, and the requirement for metal furring that performs a secondary function, a meaningful saving can be realized.

Estimated Cost Summary

Total cost per square foot	$1.59	*Total savings*
Cost reduction	13%*	Roof section
Total savings	$3500	$1.79 - 1.59 = $0.20
		Building height reduction
		$0.07 - 0.02 = $0.05
		Total $0.25

*Including savings from reduced building height.

BEFORE

	Cost/ Sq ft
Smooth surface built-up roof (common)	N/A
1 1/2" rigid insulation	$0.27
Metal deck 26 ga.	0.36
Steel joist (common)	N/A
6" furring channels	0.35
6" rock/wool bat insulation	0.46
Firecode sheet rock	0.35
Suspended acoustical ceiling (common)	N/A
TOTAL	$1.79

| UNIT | COMPONENT | FUNCTION | | | EXPLANATION | WORTH | ORIGINAL COST |
		VERB	NOUN	KIND			
Sq ft	Roofing				Item common to both designs		N/A
	Roof insulation	Control	Elements	B		0.30/2= 0.15	$0.27
		Occupy	Space	S	1/2" of space		0.01*
	Steel joists				Item common to both designs		N/A
	Rock wool bat insulation	Control	Elements	B		0.30/2= 0.15	0.46
		Occupy	Space	S	6" of building height		0.07*
	Metal deck	Support	Load	B	Item common to both designs, Dead and live load	0.30	0.36
	Metal furring channels	Support	Weight	S	Holds gypsum board & bats	----	0.35
	Gypsum board	Control	Elements	B	Fire	0.10	0.35
	Acoustical ceiling	Control	Elements	B	Item common to both designs		N/A
					VALUE	0.70	
					Cost/value =1.87/0.70=2.7*		
					TOTAL COST	$1.87/sq	ft

AFTER

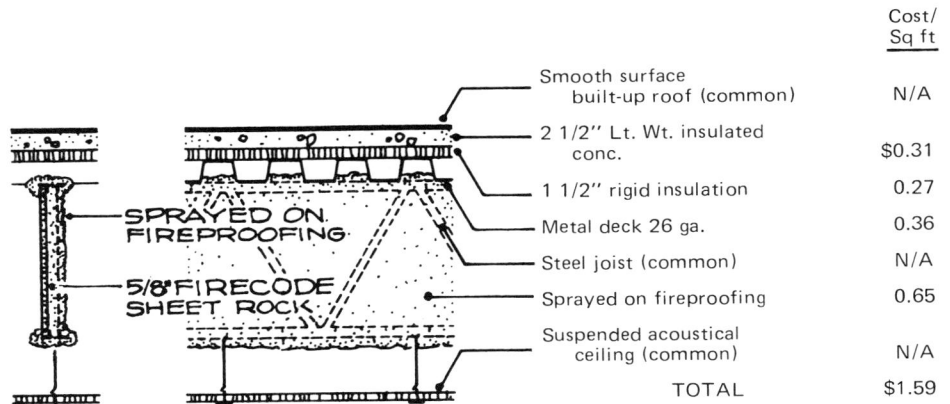

	Cost/ Sq ft
Smooth surface built-up roof (common)	N/A
2 1/2" Lt. Wt. insulated conc.	$0.31
1 1/2" rigid insulation	0.27
Metal deck 26 ga.	0.36
Steel joist (common)	N/A
Sprayed on fireproofing	0.65
Suspended acoustical ceiling (common)	N/A
TOTAL	$1.59

Labels on diagram: SPRAYED ON FIREPROOFING; 5/8" FIRECODE SHEET ROCK

EXAMPLES OF ACTUAL RESULTS 79

► **Example 3: Hidden Cost of Interest**

Description of Study

This area was isolated as a result of applying the total cost concept, which includes interest charges, to isolate unnecessary costs. Because of the recent rise in the "cost" of money, interest charges incurred by contractors assume an important status. When this requirement was originally imposed, interest rates varied from 4% to 5%. Today costs range from 9% to 10%. The savings opportunities presented can be implemented without adverse effect on building requirements.

PAYMENT SCHEDULE

**GSA REQUIREMENTS
FORM 1139**
(General Conditions)
Paragraphs 12.2 & 12.3

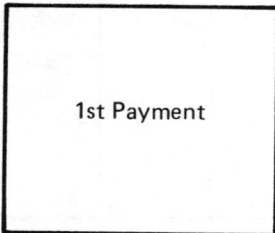

COMMERCIAL PRACTICE

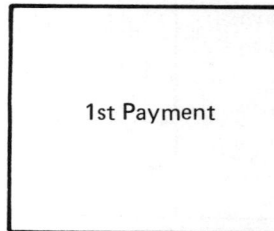

1st Payment

1st Payment

Bond & Preparatory Costs
(1-2% of Project Cost)
Prorated

Bond & Preparatory Costs
Included

SAVINGS UP TO $10,000
(Interest only)

▶ **Example 4: Effects of Standards on Cost of Building Construction**

Description of Study

This area was isolated as the result of the total cost concept used to isolate unnecessary costs. The cost of paperwork is considered part of this concept. This area is particularly sensitive to savings in a large project ($10 million) because of the recent increase in overhead costs. The study team considered the required function and sought a lower-cost alternate.

MATERIALS APPROVALS

AGENCY REQUIREMENTS
(All product specifications)

COMMERCIAL
PRACTICE

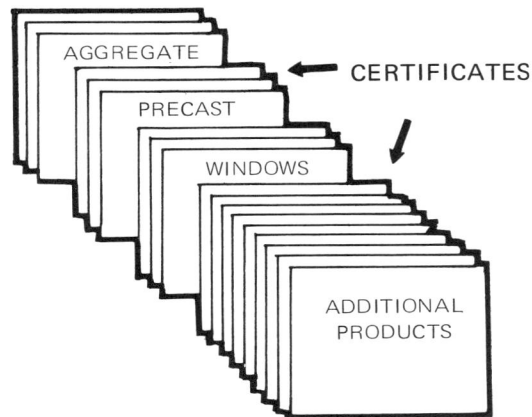

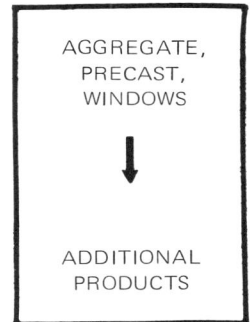

AGGREGATE

PRECAST

WINDOWS

← CERTIFICATES

ADDITIONAL
PRODUCTS

AGGREGATE,
PRECAST,
WINDOWS

↓

ADDITIONAL
PRODUCTS

In Excess Of
50 Products

Letter
Requesting Approval
Of Each Source

SAVINGS UP TO $5000

► **Example 5: Handling and Reproduction of Drawings**

Description of Study

The team isolated this area for study because of the high cost of handling and reproducing the large "F"-size prints and because of the numerous offices that were filing and storing prints.

The study determined that by using microfilm, maximizing use of half-size "C" prints for those areas where the smaller size would be adequate (e.g., material take-off), and eliminating unnecessary filing and storing, a savings of over $80,000 could be realized for the project valued at $150 million.

The information phase, which included the development of a flow chart and cost information, was the most difficult of the study.

BEFORE

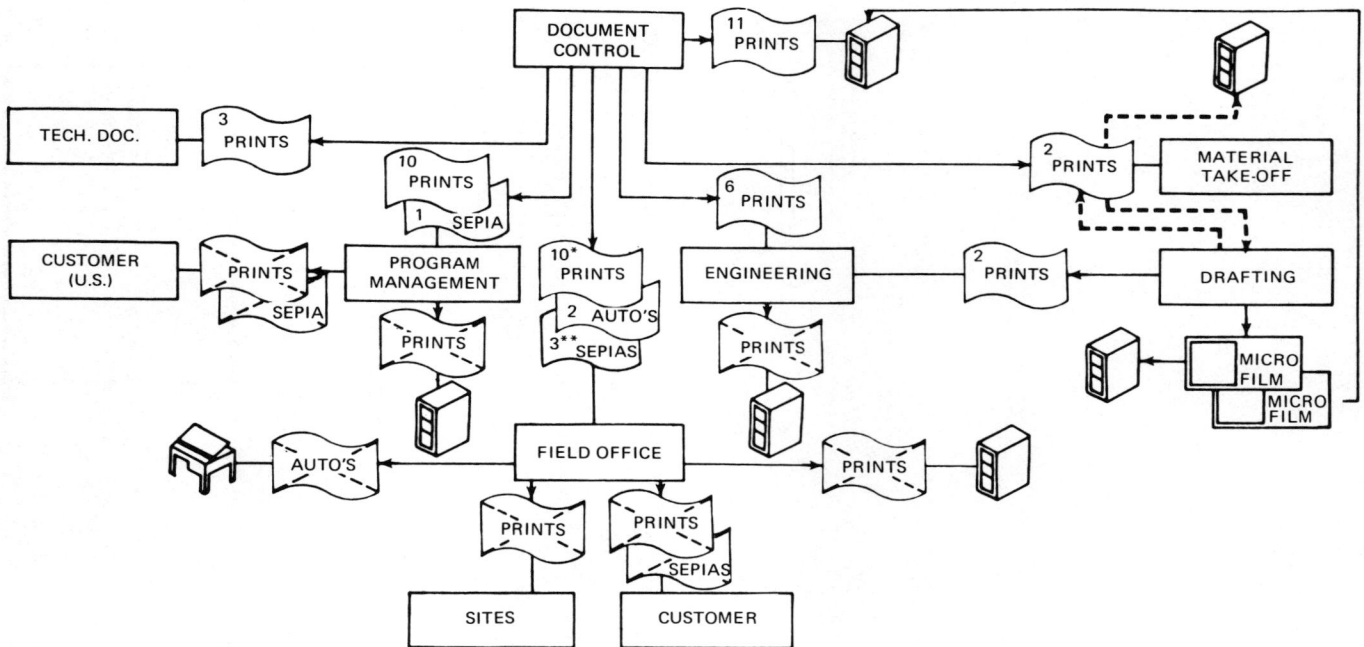

*This quantity is for installation type drawings for
  site; 80 prints are common
**For site construction drawings only.

AFTER

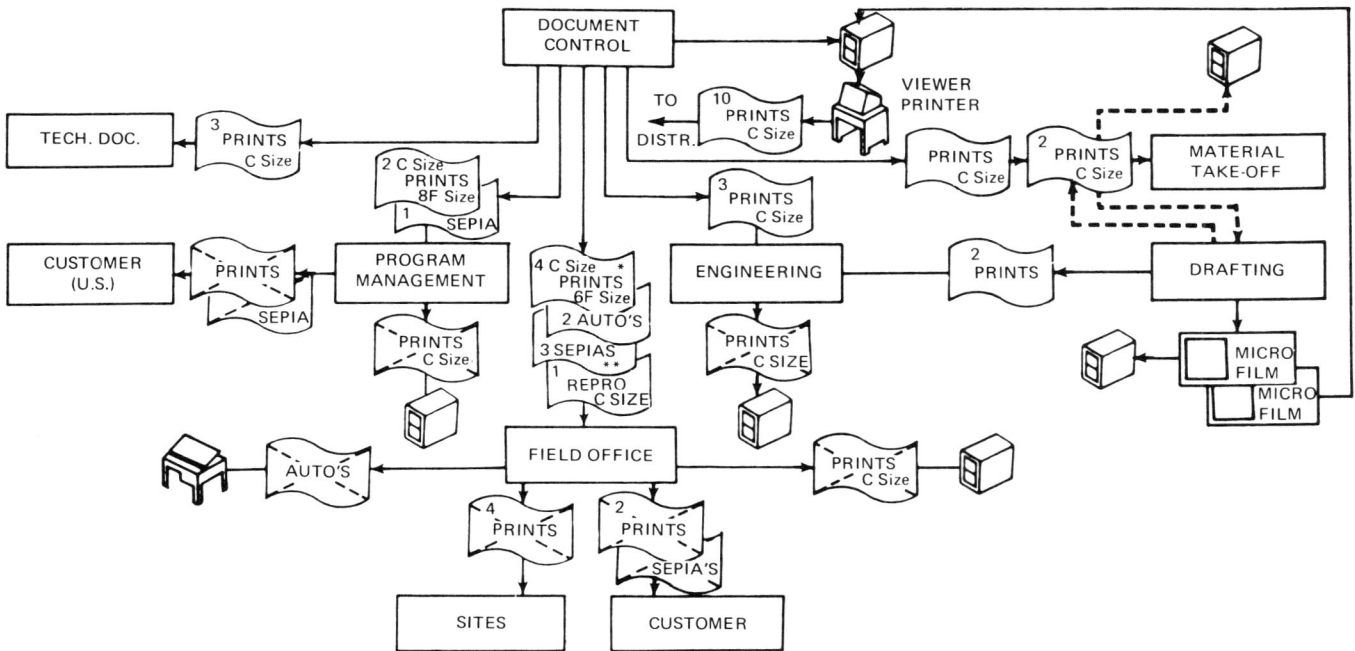

*This quantity is for installation type drawings for
  site; 80 prints are common.
**For site construction drawings only.

▶ Example 6: Fuel Storage and Distribution Area

Description of Study

The team initially isolated this area for study because of the use of two tanks and the high maintenance cost to break and repair the concrete apron when the tanks required replacement.

During the information phase the team isolated data which indicated that more significant savings could be realized in the purchase of fuel than in maintenance costs. The design was quite old, and when developed, tank truck sizes were only 5000 to 6000 gal. Newer gasoline tank trucks are sized at 8000 to 9000 gal. Full-tank loads cost $0.169/gal. with the smaller trucks and $0.151/gal. with the larger vehicles.

The team recommended that for future facilities with enough land area, the tanks be placed outside of the pavement and be increased from two to three units to take advantage of full-tank load prices.

BEFORE

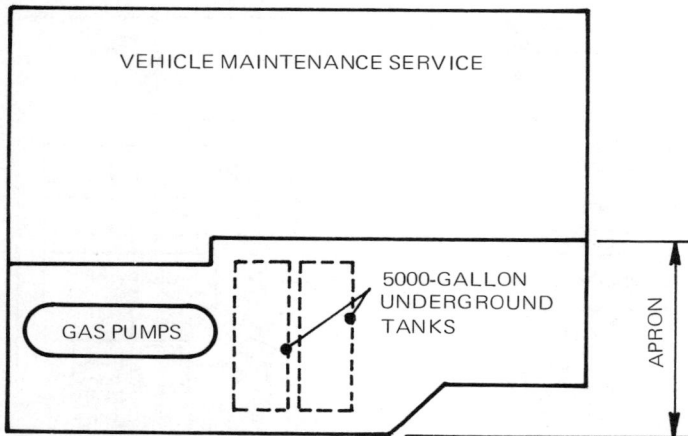

Fuel used per year ·   118 Million Gallons

Fuel purchased through central fuel service · · · · · · · · · · ·   47 Million Gallons

Fuel purchased in tank truck ·   5 Million Gallons

Fuel purchased in tank wagon ·   42 Million Gallons

COSTS OF CENTRAL FUEL SERVICE
5 Million Gallons @ $0.151	$ 725,000
42 Million Gallons @ $0.169	$7,098,000
TOTAL	**$7,823,000**

Cost Summary

Before	$7,823,000
After	7,097,000
Savings	$ 726,000

AFTER

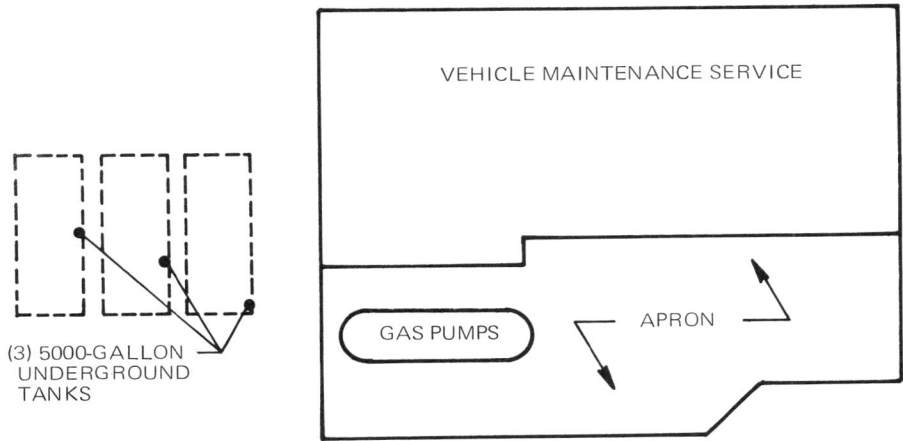

(3) 5000-GALLON UNDERGROUND TANKS

VEHICLE MAINTENANCE SERVICE

GAS PUMPS

APRON

Fuel purchased through central fuel service - - - - - - - - - - - - 47 Million Gallons

Fuel purchased in tank truck - 47 Million Gallons

COSTS
47 Million Gallons @ $0.151 $7,097,000

TOTAL   **$7,097,000**

► **Example 7: Relocation of Machine Room**

Description of Study

The machine room location was earmarked for study because of the high cost of the screen wall and vibration isolation. These items were performing only secondary functions; yet, they represented a significant bulk of costs.

In addition to developing information on the screen wall and vibration isolation, the team challenged the floor loading requirement of 100 lb/sq. ft. for a general office building.

It was also noted that the machine room was located in the penthouse because of the owner's concern about water problems with machine rooms in basements.

The team's alternate suggestion had considerable impact on total costs. The new design included special waterproofing costs for the underground construction. In addition, significant savings were realized in the foundation, structural design, and elevators due to changes in floor loading and machine room location.

BEFORE

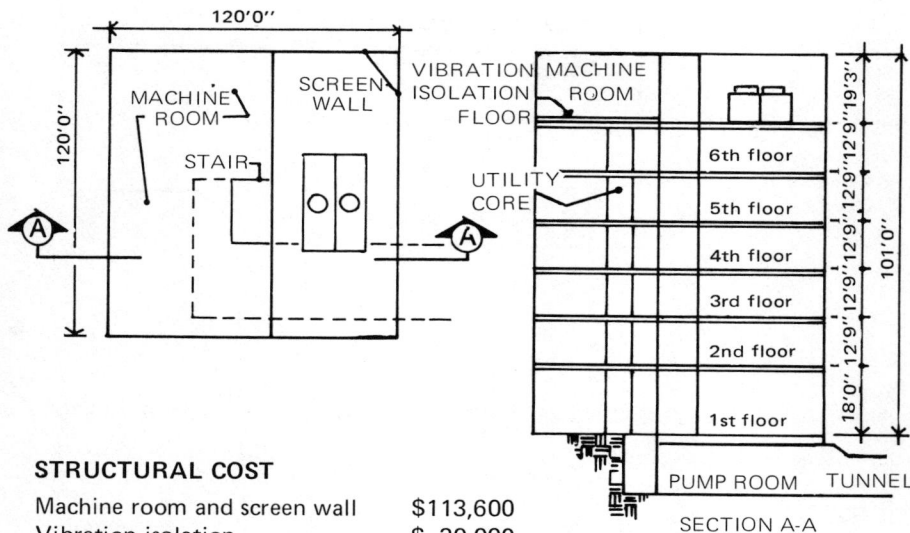

STRUCTURAL COST

Machine room and screen wall	$113,600
Vibration isolation	$ 30,000
TOTAL COST	**$143,600**

Cost Summary

Before	$143,600
After	51,900
Savings	91,700

AFTER

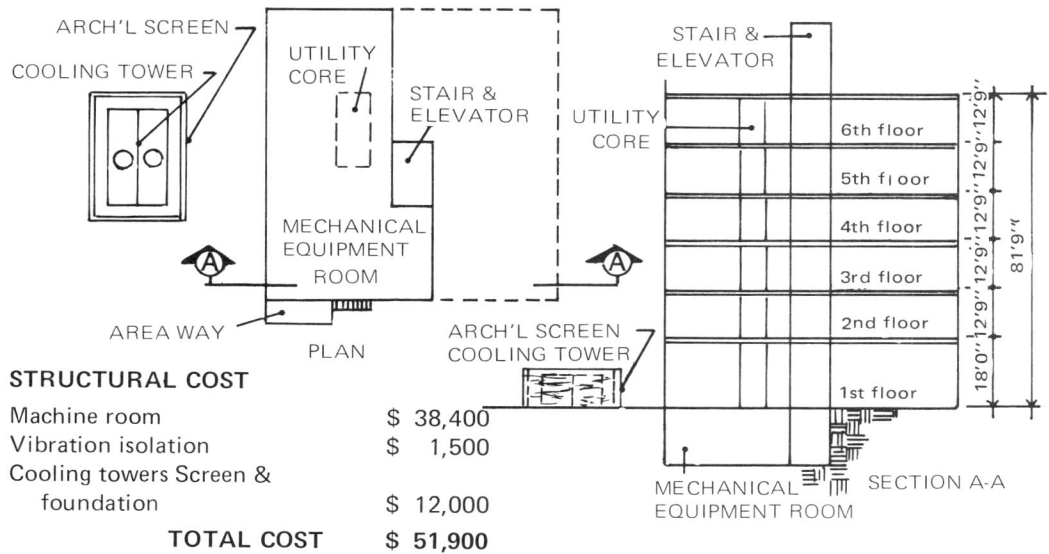

STRUCTURAL COST

Machine room	$ 38,400
Vibration isolation	$ 1,500
Cooling towers Screen & foundation	$ 12,000
TOTAL COST	**$ 51,900**

► **Example 8: Foundation Study**

Description of Study

This area was isolated primarily because of the many pile cap configurations indicated on the drawings. In addition, a change in the overall building floor loading from 100 to 75 lb./sq. ft. was to be suggested from a different study (see Example 9).

The team concentrated its effort on reducing the pile cap concrete, which is performing a secondary function. The reduced concrete would not adversely affect system requirements. By using higher capacity piles a reduction in the number of piles and size of pile caps can be realized.

BEFORE

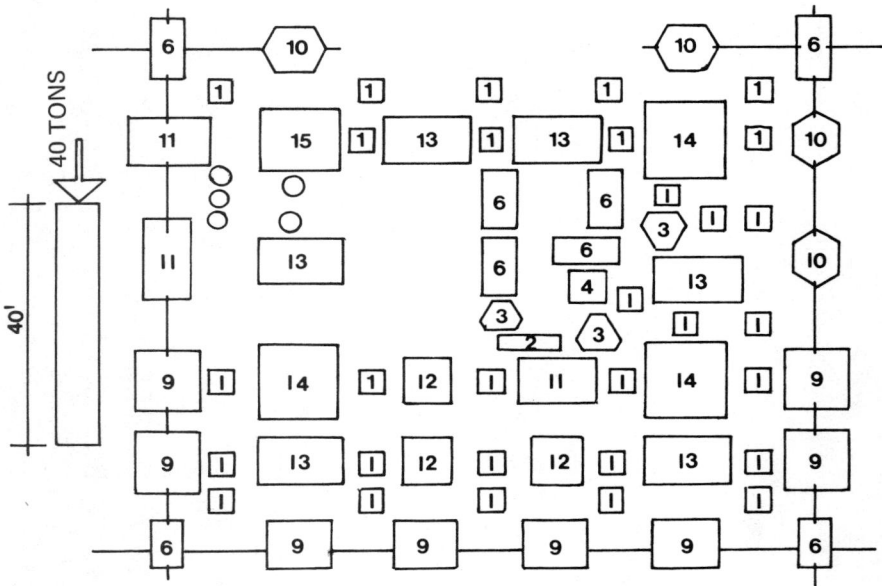

460 cast-in-place piles	$108,780
330 cu yd pile cap concrete	10,347
TOTAL COST	**$119,127**

Cost Summary

Cost of cast-in-place piles	$119,127
Cost of pressure-injected piles	79,142
Savings	$ 39,985 (33%)

AFTER

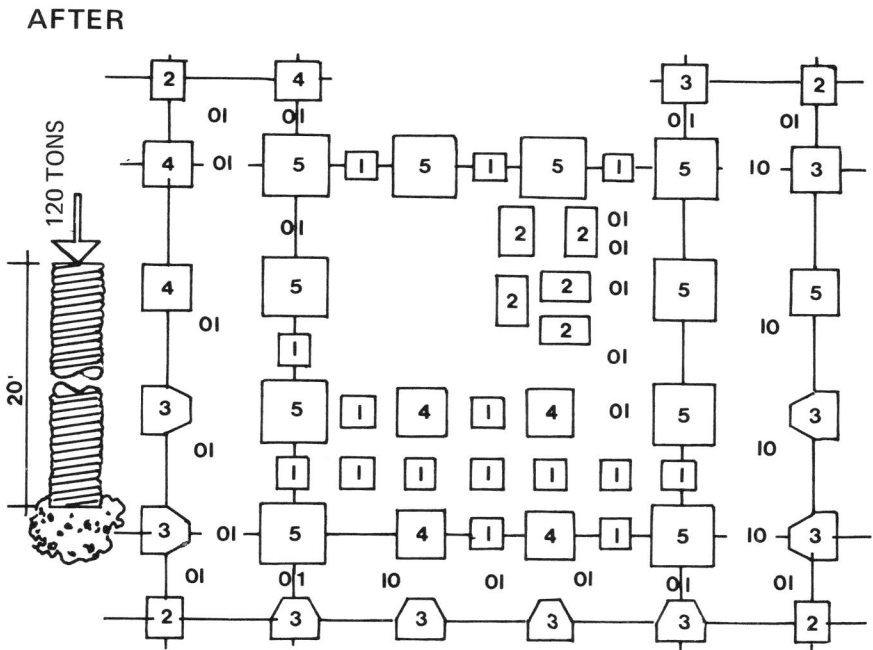

189 pressure-injected piles	$66,380
249 cu yd pile cap concrete	12,450
312 cu ft grade beams	312
TOTAL COST	**$79,142**

► **Example 9: Redesign of Concrete Columns**

Description of Study

This area was isolated because of the requirement for seven different column sizes for a seven-story building. The cost of forming these columns was thought to be quite high. The team analyzed several alternate designs and their total cost impact and decided that more money could be saved by column standardization, combined with a more realistic floor load requirement. It was also learned during the study that the owner desired movable partitions. The team questioned the costs of using movable partitions on a floor-to-floor basis in a design using many different size columns.

The team's proposal reduced the design floor load from 100 to 75 lb./sq. ft. and suggested use of 5000 lb./sq. in. concrete in lieu of 3000 lb./sq. in. concrete to standardize the size of the columns. The team felt this alternate represented the optimal solution for the owner.

Note: This example points out the need for a multidisciplinary team approach: The original design was the best structural solution but not the best for overall building cost.

Cost Summary

Cost of original column design	$74,725
Cost of proposed column design	59,954
Savings	$14,771

BEFORE

| 16 x 16 |
| 18 x 18 |
| 20 x 20 |
| 22 x 20 |
| 24 x 20 |
| 24 x 24 |
| 30 x 30 |

PENTHOUSE
6TH
5TH
4TH
3RD
2ND
1ST

AFTER

| 16 x 16 |
| 16 x 16 |
| 16 x 16 |
| 16 x 16 |
| 20 x 20 |
| 20 x 20 |
| 20 x 20 |

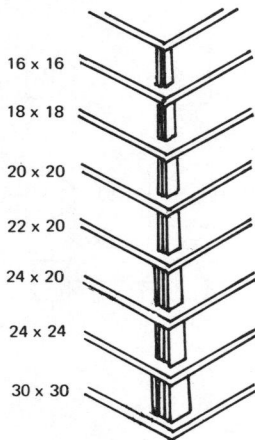

Conc strength	3000 lb/sq. in.
Axial design ld	129 k
COST	**$74,725**

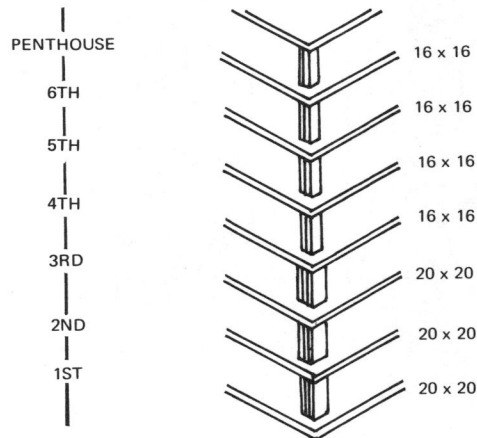

Conc strength	5000 lb/sq. in.
Axial design ld	106 k
COST	**$59,954**

▶Example 10: Buried Ducts in Floor System

Description of Study

This floor system was isolated for study because of the high cost impact the floor fill had on the building height and foundation loading. At approximately 60 lb./sq. ft. multiplied by 100,000 sq. ft. gross area, over 6,000,000 lb. of fill were added to the building foundation and structural system for a secondary function—"protect duct." In addition, the thickness of 4.5 in. for each of 6 floors added over 2 ft to the building height.

The team challenged the floor fill requirement and after substantial investigation, which included gathering test results to substantiate slab shear at dual duct junction, proposed that the duct be placed within a structural slab rather than in fill and that a larger pan size be used to effect savings in structural concrete. The owner agreed.

The total savings for this alternate was $59,650.

BEFORE

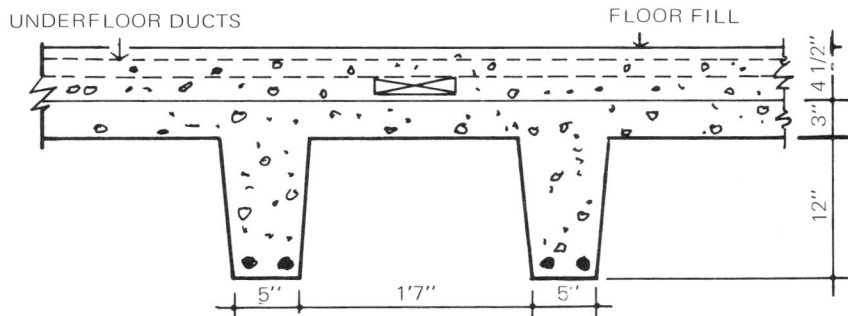

AFTER

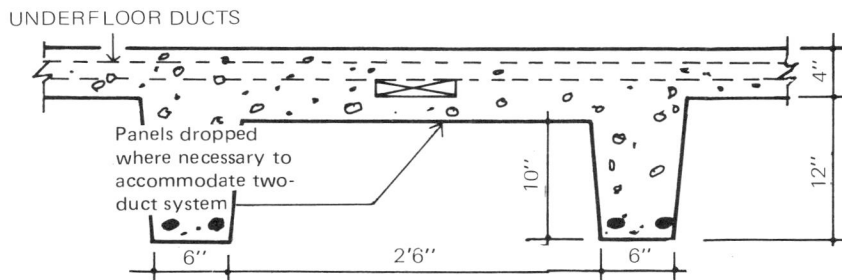

► Example 11: Elevator Study

Description of Study

The team isolated this area for study because of the speed and capacity of the elevators. The use of 200-ft./minute elevators for 60 ft. of rise was questioned because of the substantially greater cost for high-speed elevators, and because there was a question whether they would ever reach full acceleration. In addition, the use of a 5000-lb geared freight elevator was expensive for the type of facility and anticipated usage.

Investigation revealed that the director of the facility, whose office was on the top floor, demanded fast elevators. As a result, no change was made in passenger elevator speed. However, by moving the mechanical room to the basement a smaller freight elevator could be used. And although the team felt that all elevators should be hydraulically operated, only the freight elevator was changed to hydraulic to save on initial cost and reduce monthly contract maintenance costs.

Cost Summary

Cost of geared freight elevator	$45,000
Maintenance	40,000
Total	85,000
Cost of hydraulic freight elevator	24,000
Maintenance	24,000
Total	$48,000
Savings	$37,000 (16%)

BEFORE

Geared Passenger	Geared Passenger	Geared Freight
Cap 2500 lb. 200 ft/min.	Cap 2500 lb. 200 ft/min.	Cap 5000 lb. 200 ft/min.

COST	$70,000	$45,000
MAINTENANCE	$72,000	$40,000

AFTER

Geared Passenger	Geared Passenger	Hydraulic Freight
Cap 2500 lb. 200 ft/min.	Cap 2500 lb. 200 ft/min.	Cap 3500 lb. 150 ft/min.

COST	$70,000	$24,000
MAINTENANCE	$72,000	$24,000

► **Example 12: Toilet Room Layout for a Laboratory**

Description of Study

The team recommended relocating the water closets back to back and closer to existing DWV (drainage, waste, and vent) to reduce piping requirements. The revised layout of the partitions produced a 7% increase in net usable floor area.

The savings effected totaled $1300.

BEFORE

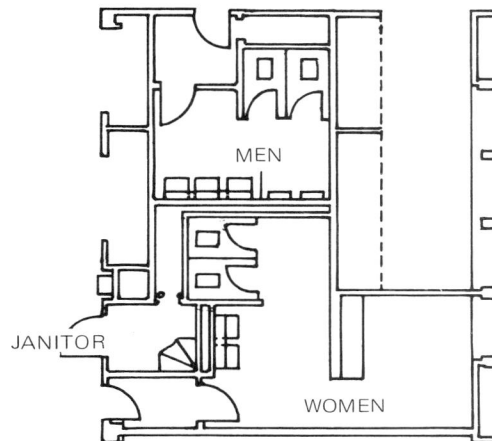

AFTER

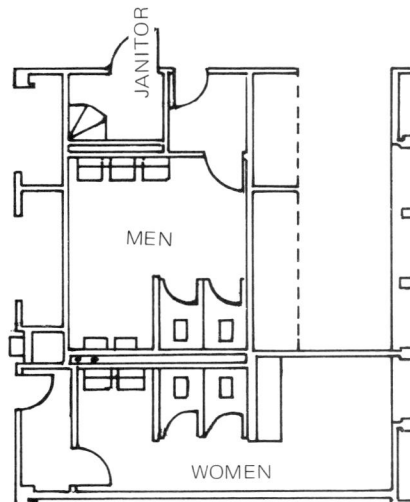

▶ Example 13: Roof Construction for One-Story Photo Laboratory

Description of Study

The original plans called for 41-in.-deep built-up girders for roof construction. The team suggested using standard rolled sections W 24 x 94, of greater weight but with lower unit cost. Further savings would be affected by reduction of 1.5 ft in building height.

The potential savings amounted to $10,000.

BEFORE

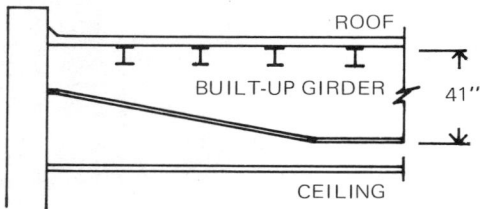

AFTER

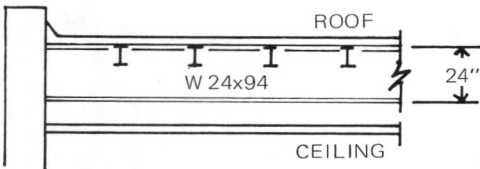

▶Example 14: Penthouse Structure

Description of Study

The VE team studying the proposed medical science building described earlier (see Figure 44) recommended changing to a steel frame, eliminating the stone coping (save $13,000); using 4-in. brick and 6-in. block for the wall (save $8,000); using painted galvanized louvers instead of Duranodic aluminum (save $12,000); and making miscellaneous other changes (save $1,700).

Cost Summary

Before	$85,340
After	50,640
Savings	$34,700 (41%)

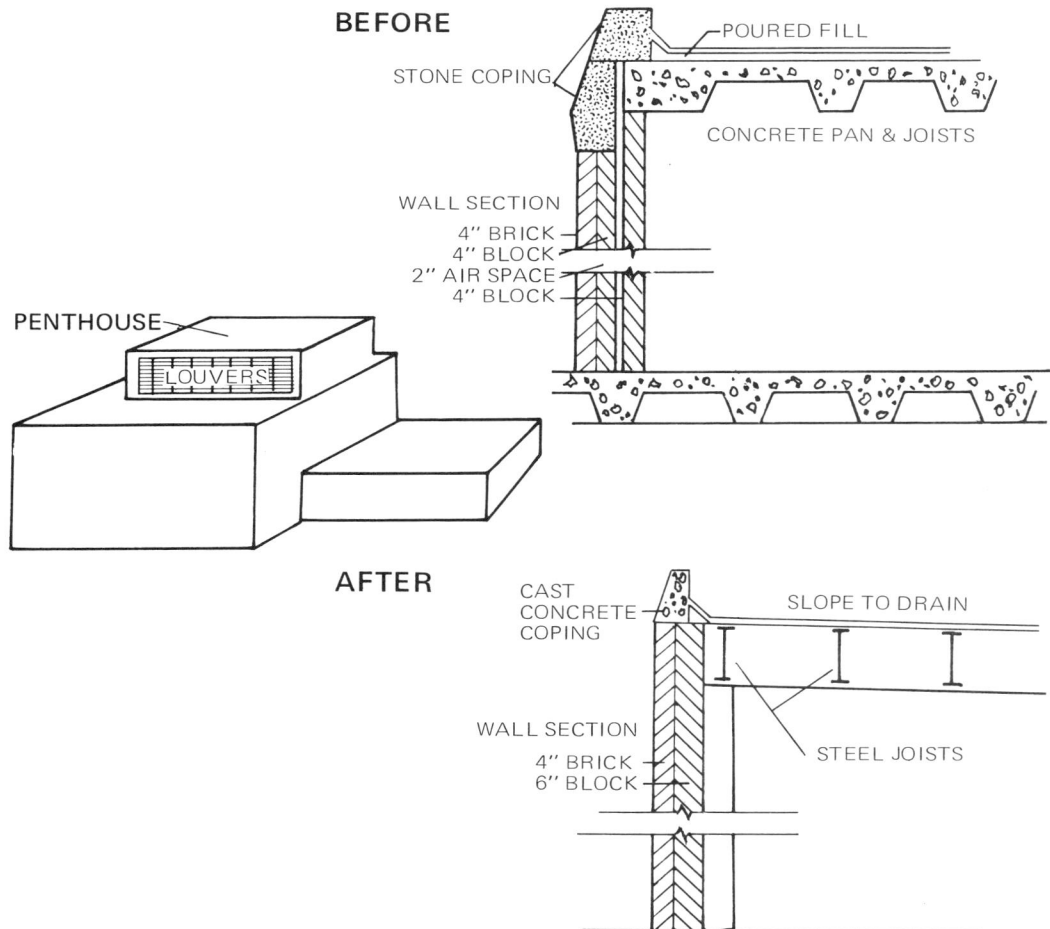

▶ Example 15: Fume Hood Exhaust System

Description of Study

For a planned medical science building (See Figure 44) the VE team recommended using a 24-guage stainless steel duct (instead of 18-gauge) to the fire damper, then changed it to terne metal. At the suggestion of maintenance people who would not change the filter due to possible contamination, it also suggested relocating the filter box from the penthouse to over the fume hood. To reduce initial cash outlay the team recommended not purchasing 25 HEPA filters at over $1200. each, and purchasing only those twelve needed for special testing. An ultraviolet lamp was added to each fume hood to kill bacteria before entering the filter.

Cost Summary

Before	$250,000
After	157,000
Savings	$ 93,000 (37%)

BEFORE

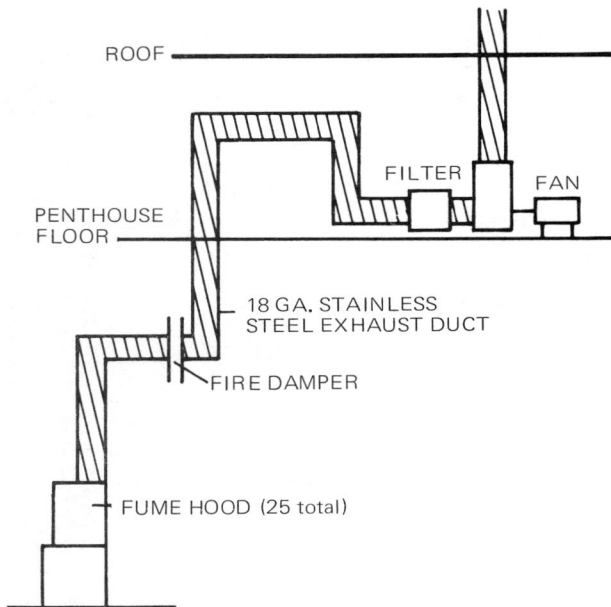

AFTER

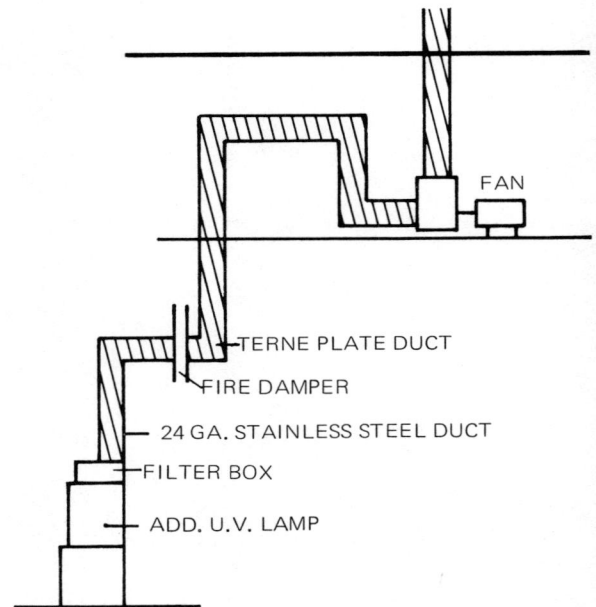

► **Example 16: Hot Water Heating System**

Description of Study

The team's proposal changed the planned medical science building's (Figure 44) original two heating systems (steam for the perimeter, hot water for reheating) to a single heating system using hot water only. This reduced the number of reheat boxes needed from 202 to about 150 by combining like exposure zones.

Cost Summary

Before	$406,000 (approx. $4.00/sq ft)
After	323,250
Savings	$ 82,750 (20%)

BEFORE

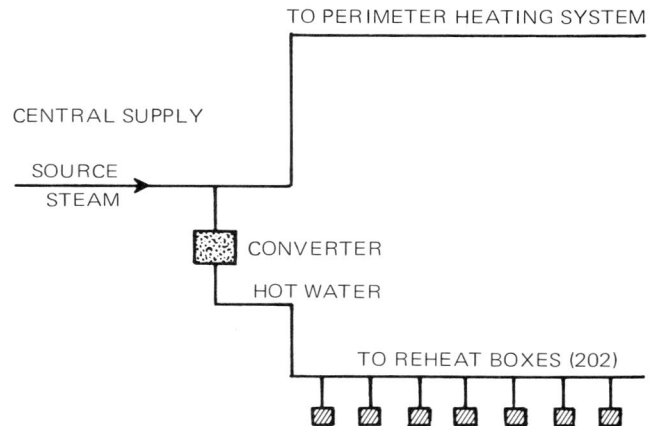

AFTER

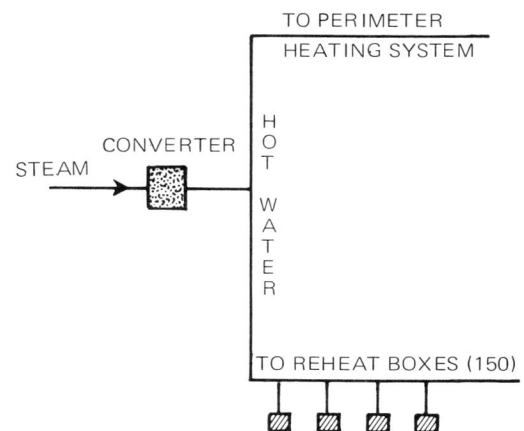

► **Example 17: Lighting Fixture Layout**

Description of Study

For the medical science building (Figure 44) the team recommended using less expensive lighting fixtures, revising the diffuser system, and reorienting the fixtures. These suggestions maintained uniformity of light over work area but reduced uniformity for passage and corner areas.

Cost Summary

Before	$ 86,560
After	43,485
Potential savings, instant	43,075 (52%)
Potential savings, life-cycle (20 yr)	200,000

BEFORE

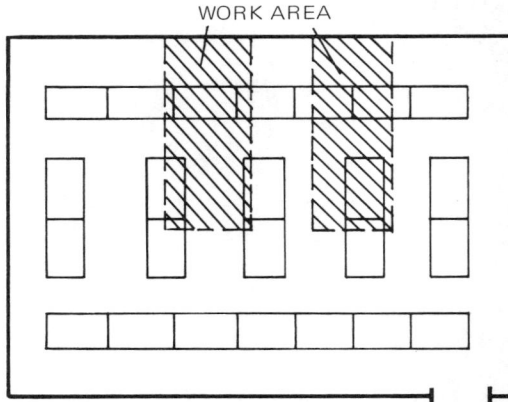

24 Fixtures   100 fc
3 Lamp wrap-
   around   Area-756 sq ft
4 Watts/
   sq ft   CU 0.5, MF 0.7

AFTER

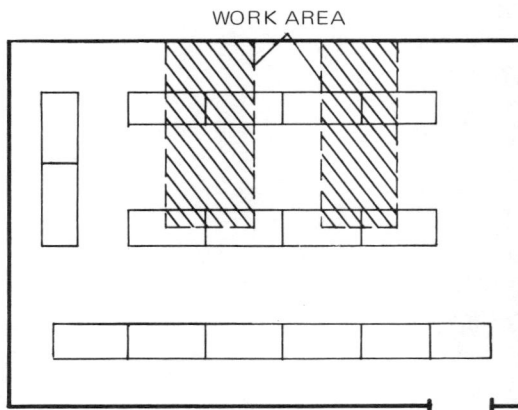

16 Fixtures   100 fc
Bare Lamp w/
   reflector   CU 1.0
3 Watts/sq ft   MF 0.8

► **Example 18: Sprinkler System**

Description of Study

For the planned telephone exchange facility described earlier (Figure 46) the VE team suggested revising the number and type of sprinkler heads and deleting the planting area in the parking lot.

These alterations effected a saving of $10,700, a 30% cost reduction.

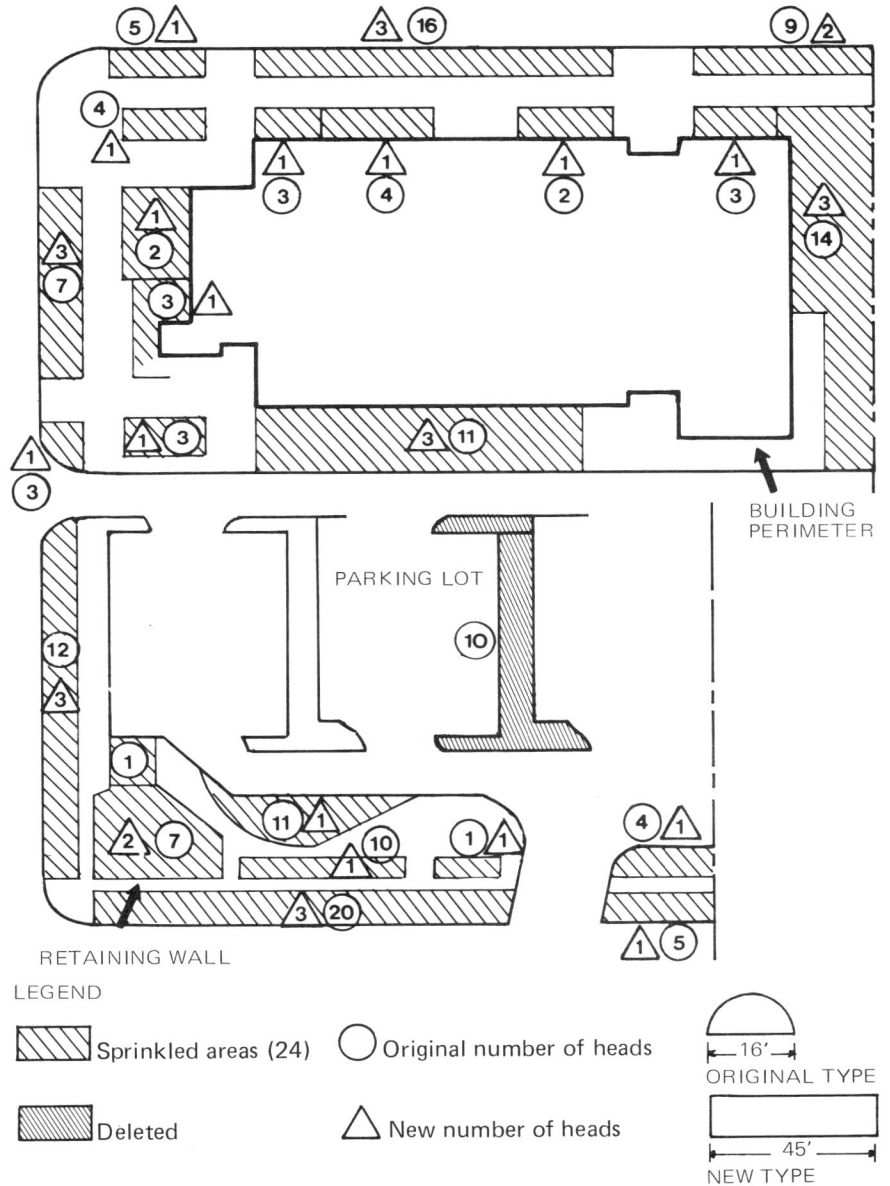

BUILDING PERIMETER

PARKING LOT

RETAINING WALL

LEGEND

⬚ Sprinkled areas (24)   ◯ Original number of heads   ⌂ ORIGINAL TYPE ⊢16'⊣

▨ Deleted   △ New number of heads   ▭ NEW TYPE ⊢45'⊣

► **Example 19: Roofing Design**

Description of Study

For the telephone exchange building (Figure 46) the team recommended the use of tapered insulation board in lieu of lightweight concrete roof fill. It also suggested deleting the parapet and using gutters and exterior roof drains. Planned future expansion costs would be substantially reduced as well as initial costs.

Cost Summary

Before	$51,600
After	41,600
Savings	$10,000 (20%)

BEFORE

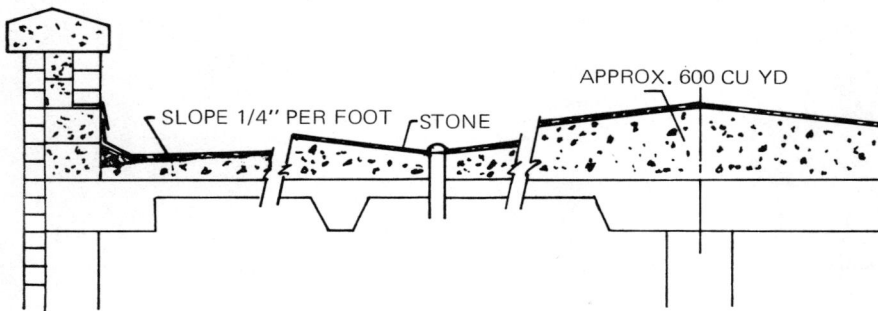

AFTER

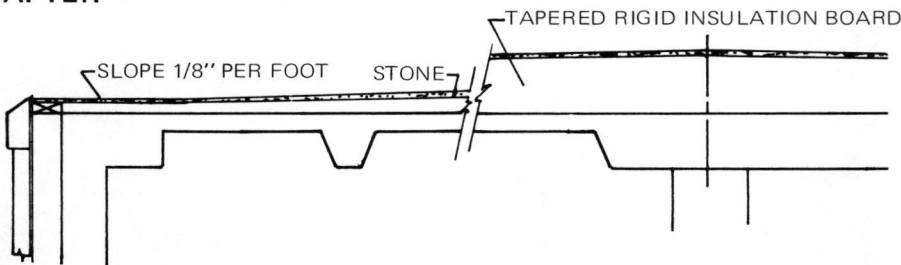

▶ **Example 20: Slab Floor System**

Description of Study

The team studying the telephone exchange facility (Figure 46) recommended changing the original two-way pan slab with topping to a one-way slab with a shallow beam (topping requirement eliminated), to avoid the high cost of topping unnecessary with present technology.

Cost Summary

Before	$225,000
After	201,000
Savings	$ 24,000 (10%)

BEFORE

Cost/sq ft = $4.68
Total sq ft = 48,000
TOTAL COST $225,000

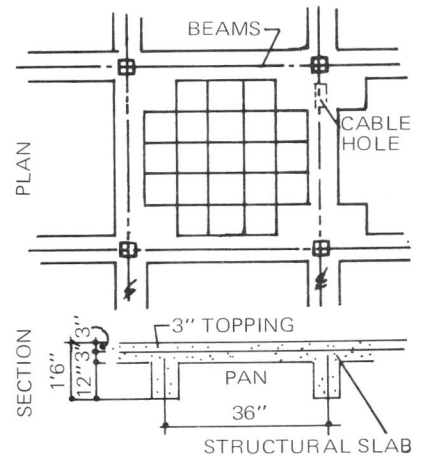

AFTER

Cost/sq ft = $41.8
Total sq ft = 48,000
TOTAL COST = $201,000

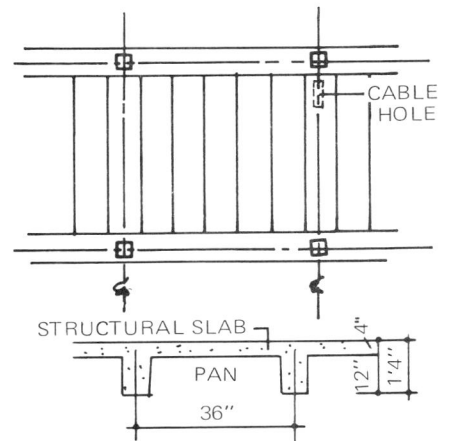

▶ **Example 21: Steam Heating System**

Description of Study

For the telephone exchange facility (Figure 46) the team recommended omitting hot water baseboard heating and hot water coils, substituting steam coils and electric baseboard radiators where required. The savings in the heating system were even more significant since equipment-generated heat in the building exceeded the heat required for even the coldest temperature.

The savings totaled $45,000.

BEFORE

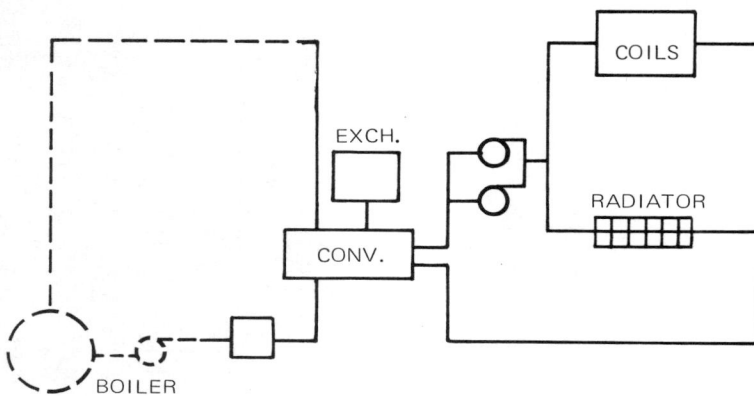

AFTER

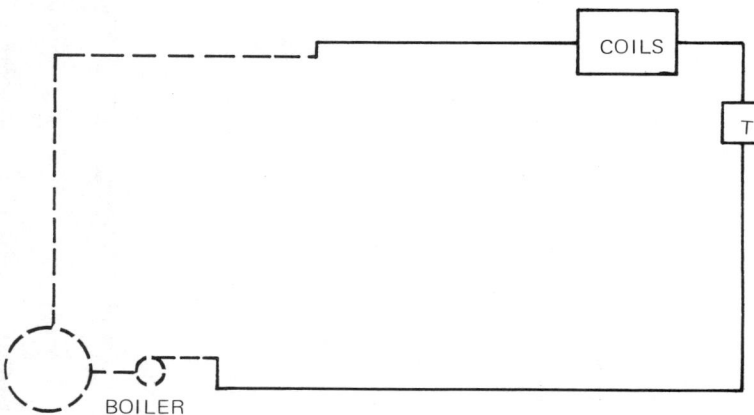

► **Example 22: Lighting Plan for the Reduction of Power and Air-Conditioning Costs**

Description of Study

The team studying the telephone exchange facility plans (Figure 46) recommended use of 30 fc maximum, thus reducing number of fixtures and impact on air-conditioning load.

A 15% reduction was affected in installation costs, and further equipment savings accrued from air-conditioning load reduction. The building equipment was fully automated requiring low-level occupancy.

Cost Summary

Before	$5,975
After	4,900
Savings	1,075
A/C load reduction ≈ 1 ton	1,400
Total savings	$2,475 per floor

EXISTING

Existing Fixtures
31 Fluorescent 8' = 6200W
58 Fluorescent 4' = 5800W
89 TOTAL
   FIXTURES   = 12,000W

PROPOSED

Proposed Fixtures
69 Fluorescent 4' = 6900W
13 Incandescent = 1300W
82 TOTAL
   FIXTURES   = 8200W

LEGEND
-•-   Incandescent Lights
▬   4' Fluorescent Lights
╋   8' Fluorescent Lights

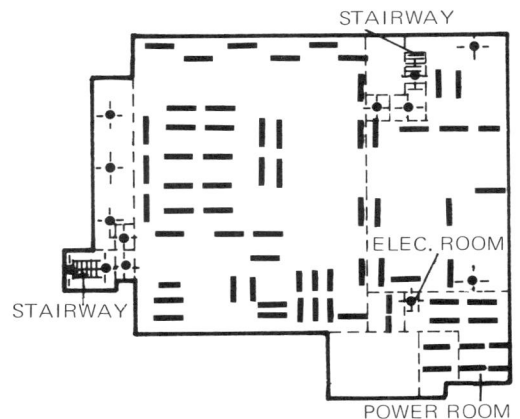

▶ Example 23: Alternating-Current Switchboard Revision

For the proposed telephone exchange facility (Figure 46) the team recommended the use of switches with current-limiting fuses in lieu of circuit breakers; it also suggested changes in the cabinet depth.

BEFORE

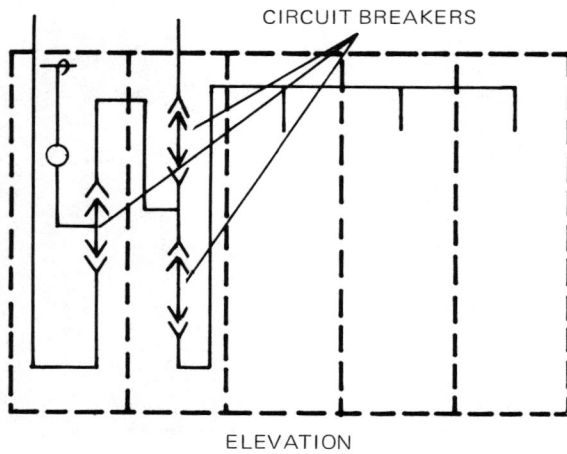

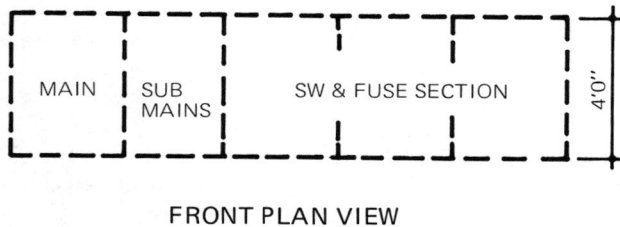

FRONT PLAN VIEW

Cost Summary
Before $22,860
After   14,800
Savings $ 8,060 (35%)

AFTER

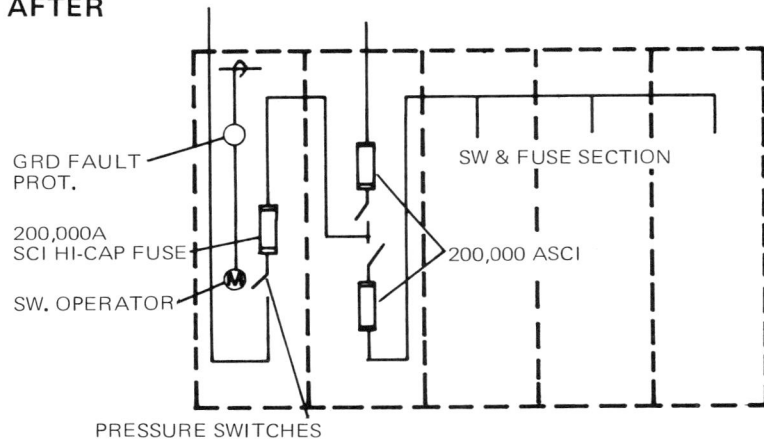

GRD FAULT PROT.

200,000A SCI HI-CAP FUSE

SW. OPERATOR

PRESSURE SWITCHES

SW & FUSE SECTION

200,000 ASCI

ELEVATION

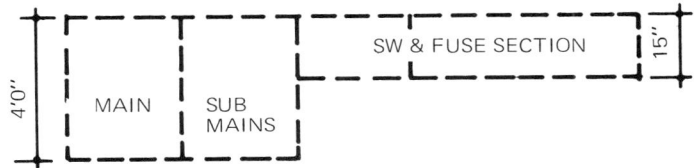

4'0"

MAIN

SUB MAINS

SW & FUSE SECTION

15"

FRONT PLAN VIEW

► **Example 24: Clean Room Duct Layout**

Description of Study

The team recommended changing the conventional duct run with individual filters at each diffuser to a factory-made duct run with a continuous filter arrangement.

Cost Summary

Before	$210,800
After	66,150
Savings (on duct work alone)	$144,650

BEFORE TYPICAL DUCT SECTION (24 REQUIRED)

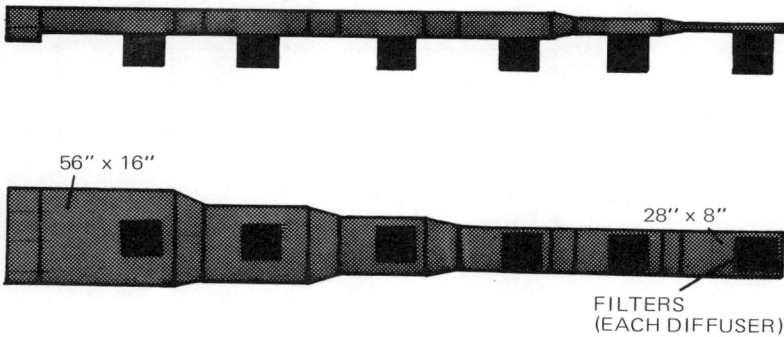

56″ x 16″

28″ x 8″

FILTERS (EACH DIFFUSER)

AFTER

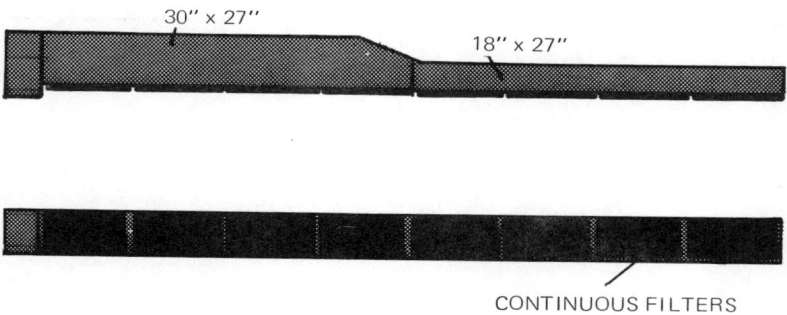

30″ x 27″

18″ x 27″

CONTINUOUS FILTERS

VE ORGANIZATION IN THE CONSTRUCTION INDUSTRY

Industry Characteristics

Presently, the construction industry, although practicing many of the techniques of VE (e.g., economic analysis studies by structural engineers), basically uses a compartmentalized approach to the design of facilities. As a result, each division or unit is responsible for issuing, reviewing, and updating the criteria and requirements of its own specialty. This approach tends to optimize subsystem performance and costs without proper consideration to end-system function and total costs. For example, structural engineers seek to optimize structural costs; yet, what direct relationship does this approach have to the lowest total system costs? In other words, it is possible to have an optimum structural design, but the effect on the total building (such as added height) can cause higher overall costs (see Example 13).

A VE program will augment existing efforts and take a fresh look at design by using a team approach in a creative atmosphere. This additional effort will cut across organizational lines and challenge selected high-cost items regardless of areas or disciplines involved. A significant potential for VE does exist through use of a team approach during the design review cycle of major facilities. In addition, a similar potential exists for a continuing program of critical appraisal of existing standards and building requirements.

Organization

The prerequisite for establishing a VE program is the strong desire by the firm's top management to allocate the necessary resources, both in effort and in funds. As a rough guide, any organization responsible for over $15 million in expenditures should consider establishing a full-time program. Smaller firms should limit their VE investment to a part-time program. For example, a VE effort should be focused on key projects which would have impact across the total organizational program. Use of an experienced VE consultant for these concentrated efforts is recommended. This approach will aid in creating a greater cost effectiveness on the decisions being made.

It is recommended that VE be assigned as a staff function of the highest level of management. As a staff function, it would have the following advantages:

1. Be able to cross organizational lines.
2. Be able to get guidance from management and administrative assistance.
3. Have freedom of operation. Items outside design (i.e., contracting) could be challenged, and the pressure of design deadlines would not be present.
4. Have access to funds for proper development and evaluation of VECPs.
5. Have excellent communications with sources of new materials and methods, and because of ability to cross organizational lines, a crossfeed of information can be established.
6. Be in a good position to monitor the performances of contractors participating in VE incentive contracts.

Most VE personnel believe that locating the VE position in the design area is not a good idea. Locating the VE function under the same supervisor having his work reviewed tends to be restrictive. The staff position selected should not be too removed from the principal areas of work, i.e., design and review functions.

The first recommendation for organizations establishing a VE program is to recruit a supervisor. A position description outlining the principal duties and responsibilities of a value engineer is found in Appendix F. Next, a training program must be initiated to inform key management officials of the purpose and functions of the program. These management briefings should be conducted by the VE supervisor.

The training program should include workshop seminars during the first year, using active projects as class problems. Outside consultants should assist in the initial training. Every effort should be made to have personnel from different disciplines participate in the training seminars.

Staff members whose decisions affect project costs must receive proper VE indoctrination. A 1-day indoctrination is normally sufficient for management personnel. However, personnel who will become actively engaged in VE studies require an additional 5 days of training by qualified instructors.

The training seminars will accomplish a number of beneficial things. Interdisciplinary communication within the organization will improve. The total cost impact of decisions will be better defined—and increased profits or greater cost effectiveness gradually realized. Appendix G presents the scope of work of a typical VE workshop seminar.

Implementing Incentive Clauses

It is recommended that incentive clauses not be included in construction contracts until after the initial indoctrination of key management personnel and establishment of a functioning VE staff.

Only after these criteria are met (normally within 1 year), should incentive clauses, similar to those now used by the Department of Defense construction agencies, be placed in construction contracts exceeding $100,000 (see Appendix A). After award, the successful contractor should be briefed by the VE staff on the organization's intent and attitude regarding these clauses.

All VECPs submitted by the contractor should be coordinated through the VE staff. The VE staff will maintain records of proposals and assist in their expeditious processing. After acceptance of a VECP by management, the VE staff should be responsible for seeing that required changes are implemented on future projects.

Public Relations

In addition to the above actions, it is recommended that the VE staff develop means to publicize the program properly. Certificates of merit should be awarded to worthy individuals contributing to the program. Newsletters should be circulated to chiefs of the company's operating branch and to key office personnel. Articles describing a worthwhile achievement should be written for publication in leading journals. It is imperative to keep the program dynamic and individuals aware of the existence of the VE effort.

Management Controls

Reporting system. In order for management to justify the expenditure of funds for a VE effort, they must have some unbiased determination of results. It is recommended that a periodic report be established, outlining significant actions and accomplishments. The report should be submitted to the management director. A reporting system including format, desired information, and the basis for determining savings should be developed.

Audit system. An audit system should be established to monitor and evaluate the actions and accomplishments set forth in the reporting system. Qualified personnel from outside the VE staff, in cooperation with staff members from finance and administration, should conduct the audit.

Goal assignment. In order to motivate personnel, it is recommended, whenever feasible, that goals and/or any other system be established to stimulate active participation.

Feedback system. The VE staff should establish a feedback system throughout the organization for all VE changes. The results of a survey of a government agency having various field offices, showed that approximately 30% of the savings generated at each office resulted from information received from the feedback system.

A practical example of how to implement a program is contained in some recent General Services Administration actions. In Appendix H are found VE-related clauses that will be included in all major General Services Administration contracts. Part I lists requirements for the construction management contract, and Part II outlines the responsibilities of the architect-engineer when a construction manager is employed. These provisions result in an organized VE effort with an allocation of both time and money in a *positive additional* effort to reduce costs.

Guidelines for Potential Savings

Based on previous results and actual application of VE to specific projects, the following general estimates of savings potential are made.

On total budget. For large corporations, using results attained in the Department of Defense agencies and large industrial firms such as Thompson-Ramo-Wooldridge, Minneapolis-Honeywell, Joy Manufacturing, and General Electric, a target savings goal of 1% to 3% of total budget is reasonable.

On a large project. Using results attained through actual practice, the savings vary with the nature of the design. For large projects where costs are estimated over budgeted amounts, up to 20% reduction in total costs is possible. For designs developed under circumstances where costs appear under control, a 5% to 10% reduction in the initial construction contract and a similar benefit in the life-cycle cost are reasonable.

On selected high-cost areas within a project. Once an area has been isolated as a poor-value area, reductions of from 10% to 50% are possible. For a long-range program involving a large number of items a 30% reduction should be attained as an average.

By use of VE contract incentive provisions. On an overall program involving expenditures of over $100 million, the savings possible from contractor-initiated proposals of approximately 0.5% is a target goal having a historical base. For selected projects, a 1% to 5% reduction in initial costs is possible.

Table 6 summarizes the program savings of an agency constructing approximately $100 million annually.

PROJECT	COST OF STUDY ($)	PROGRAM COST ESTIMATE	ORIGINAL IDEA LISTING		ADOPTED FOR INCLUSION IN FINAL DESIGN	
			NO.	SAVINGS ($)	NO.	SAVINGS ($)
Computer facility	$ 7,500	$ 500,000	17	$ 92,000	17	$ 85,000
Photographic Lab	8,500	500,000	25	67,000	4	32,000
Instrumentation Lab	10,000	900,000	41	174,000	6	73,000
Office building*	10,000	2,600,000	32	486,000	8	65,000
Office building	20,000+	3,600,000	43	907,000	14	490,000
Computer facility*	10,000+	2,050,000	44	346,000	8	20,000
Chemical Lab*	8,000	1,525,000	29	183,700	6	38,600
Computer facility	9,500+	1,000,000	27	150,000	10	50,000
Spacecraft bldg.	10,000	4,500,000	35	400,000	17	230,000
Visitor's center	12,000+	2,500,000	52	150,000	27	53,000
TOTAL	$105,500	$19,675,000		$2,955,700		$1,136,600

Total cost for VE studies including management briefings: $100,000+

*Review limited to expanded idea listing (no in-depth study).
+Includes detailed cost estimate.

FIGURE 52. Saving resulting from VE studies in $100 million annual construction program.

Guidelines for Program Costs

For construction programs ranging up to $15 million annually, the allocation of funds for a VE program should range from 0.1% to 0.5%.

For construction programs greater than $15 million but less than $100 million, a program expenditure of from 0.05% to 0.3% should prove worthwhile. A minimum expenditure of approximately $50,000 is recommended.

For large construction programs over $100 million, program costs of not less than $100,000 should be budgeted, with a maximum percentage of approximately 0.1% returning optimum results.

As a guide for management to measure the VE efforts and to monitor and forecast budgets, the VE program costs should be varied with the results obtained. After the initial year of operations, management should evaluate the program results and raise or lower the VE budget in accordance with the following guide:

1. Programs up to $15 million: Savings that result should range from 3 to 10 times the cost of the VE effort.
2. Programs from $15 million to $100 million: Savings that result should vary from 5 to 15 times the cost of the VE effort.
3. Programs over $100 million: Savings should vary from 10 to 20 times the cost of the VE effort.

If results are on the high side of the scale, additional funds for the VE budget should be considered. If results are low or outside the range, reductions are warranted. In this manner management can avoid an overkill or underkill of the VE program.

FUTURE PROSPECTS

Experience to date has indicated that the VE approach should be integrated with the overall objectives of an organization. That is, application of the methodology or achieving dollar reductions should not be the only objective of the VE program. Its purpose should be to serve a useful function for management by improving the position of the organization. This requires coordination with other areas within an organization, such as data retrieval systems* and cost control and value analysis systems.

Data Retrieval Systems

Vital data, such as unit costs, manpower estimates, and the job scheduling information must be readily available to the VE program. Conversely, the VE staff can provide input for updating of costs and data on new materials and methods. This interchange of information will serve to enrich the store of useful knowledge, generate savings, and improve the operations of the entire organization.

Cost Control and Value Analysis System

An idealized overall approach to combine VE with cost control in order to increase its effectiveness is the system outlined in Figure 52. The system illustrates that for a comprehensive approach VE should be combined with the cost reviews and design reviews which are normally scheduled for a facility. For example:

1. The initial VE in-depth study should be conducted at the same time as the cost and design reviews of the tentative (preliminary) architect and engineer contract. This VE study would be primarily oriented toward an analysis of the major design decisions. A building cost model should be developed and compared with cost information supplied by estimators.

*Special efforts are required to gather and collect valid cost information for maintenance and operations. Currently, cost data are unavailable and unreliable. This is disturbing since these costs represent over 50% of the cost of ownership.

2. Next, VE monitoring should occur during preparation of early working drawings, and an in-depth VE study should be performed on the intermediate architect and engineer contract. This VE review should be primarily oriented to assure that previously approved VEPs have been incorporated and to analyze the major system components and the materials and methods selected. The focus of the effort will be governed by the progress estimate. If the progress estimate is higher than the budgeted amount, efforts should be oriented toward cost-reducing actions involving high initial cost reduction. If the estimate is within budget, efforts should be focused on life-cycle cost savings.

3. A pre-bid VE study is recommended at the time of submittal of final drawings and specifications. This study should be primarily oriented toward materials and methods outlined on plans and specifications. The depth and nature of the effort will again depend on the pre-bid estimate. For example, if large cost problems are indicated, VE efforts may be directed to new areas even if significant design changes are involved. If pre-bid estimates and budget are compatible, efforts would be focused on isolating savings having small change impact. These changes should improve life-cycle costs or increase the change order contingency.

Value Engineering in Construction Management

In the introduction it was mentioned that many owners and agencies are turning to Construction Management (CM) to improve the cost-effectiveness of their expenditures. VE provides the CM concept with a valuable tool to manage costs. A VE effort of from 5-10% of the total cost fees will provide the owner with a positive effort directed toward both initial and life cycle cost optimization. Experience has shown VE to be an effective methodology for the CM.

As for the scope of VE services, attached as Appendix J is the recommended approach. This approach was developed from application on over fifty projects totalling over 500 million dollars.

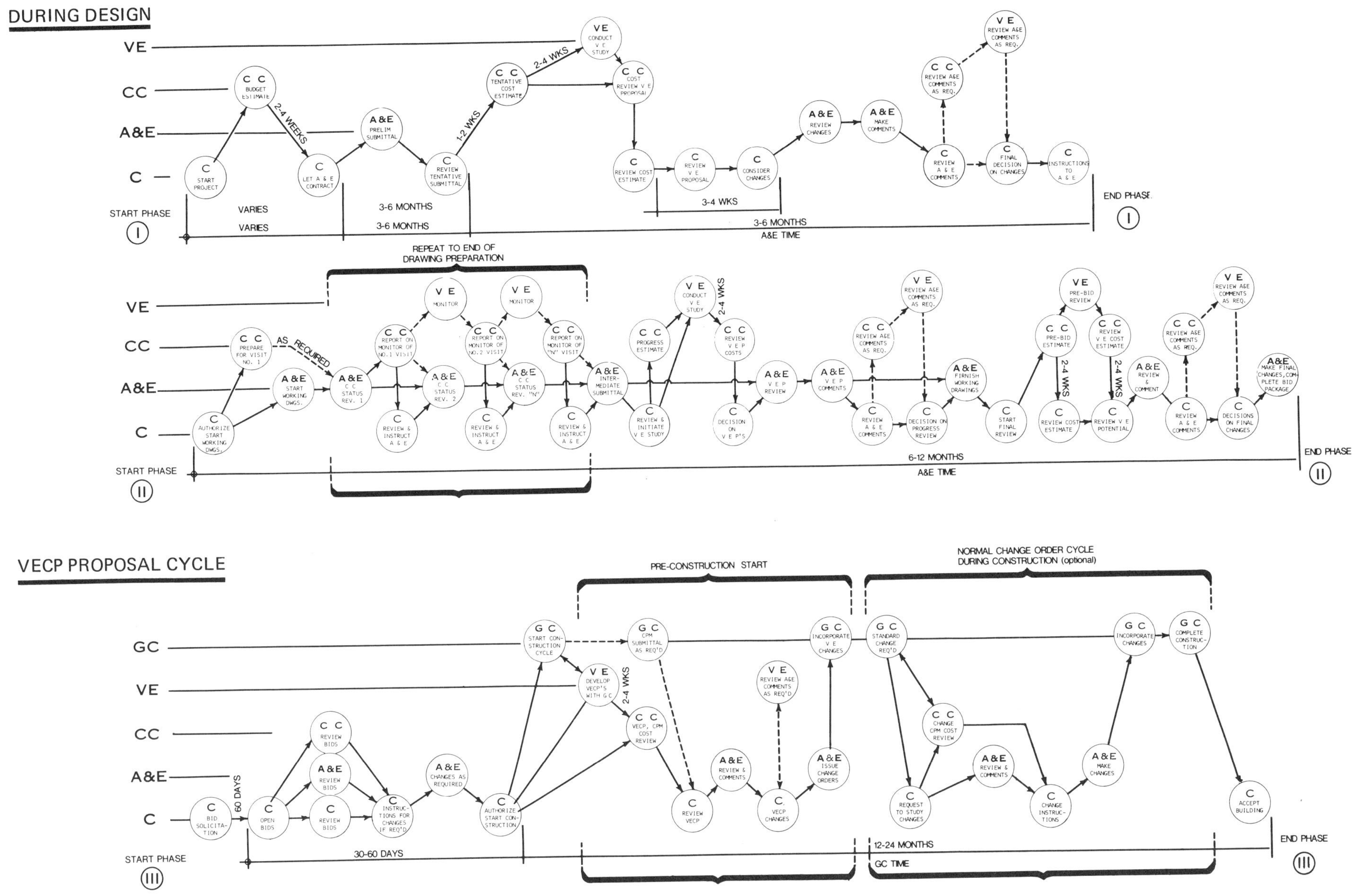

FIGURE 53. Cost and value monitoring system for major facilities. This system can be varied to fit actual project time. C, government or client; VE, value engineering, in-house or consultant; CC, cost control, in-house or consultant; A&E, architect and engineer; GC, general contractor; VEP, value engineering proposal; VECP, value engineering change proposal.

CONCLUSION

The results achieved to date demonstrate conclusively that a substantial savings potential exists through the application of VE techniques in the construction industry.

The best place for their future application is the place where the most funds are being expended. It is at the city and county government level that the greatest untouched area remains. The increasing cost of money and of construction work leaves these areas badly in need of assistance in cost and value control.

The private sector is the area of next greatest potential. With the profit incentive, the extra savings possible through the use of VE can greatly enhance profit. For example, if a contractor can share $100,000 through participation in an incentive contract, this amount is *clear* profit. To realize a profit of $100,000 under ordinary conditions, the contractor would have to complete a project of $5 million to $10 million; yet, for a small effort in developing VE change proposals, he can realize the same net profit on a smaller project. The same analogy applies to an in-house program.

As a fringe benefit to a private organization, the VE approach (which is basically a problem-solving methodology) can be utilized by management to evaluate other areas. Cost cutting does not have to be the basic parameter of a study; saving time, improving aesthetics, reducing labor, eliminating critical materials can be substituted as the objective. Under the leadership of enlightened management, a formal VE program can:

1. Increase profits or savings
2. Provide a positive means for future improvements
3. Improve the image of an organization
4. Improve internal operations and communications.

These benefits are hard to discount. They can be achieved by utilizing to the fullest extent the basic material presented in this guide.

APPENDICES

56. VALUE ENGINEERING INCENTIVE (1967 JUN)
(The following clause is applicable if this contract is in excess of $100,000)

(a) (1) This clause applies to those cost reduction proposals initiated and developed by the Contractor for changing the drawings, designs, specifications, or other requirements of this contract. This clause does not, however, apply to any such proposal unless it is identified by the Contractor, at the time of its submission to the Contracting Officer, as a proposal submitted pursuant to this clause. Furthermore, if this contract also contains a "Value Engineering Program Requirement" clause, this clause applies to any given value engineering change proposal only to the extent the Contracting Officer affirmatively determines that it resulted from value engineering efforts clearly outside the scope of the program requirement; to the extent the Contracting Officer does not affirmatively so determine, the proposal shall be considered for all purposes as having been submitted pursuant to the Value Engineering Program Requirement clause, even if it was purportedly submitted pursuant to this clause.

(2) The cost reduction proposals contemplated are those that:

    (i) would require, in order to be applied to this contract, a change to this contract; and

    (ii) would result in savings to the Government by providing a decrease in the cost of performance of this contract, without impairing any of the items' essential functions and characteristics such as service life, reliability, economy of operation, ease of maintenance, and necessary standardized features.

(b) As a minimum, the following information shall be submitted by the Contractor with each proposal:

    (i) a description of the difference between the existing contract requirement and the proposed change, and the comparative advantages and disadvantages of each;

    (ii) an itemization of the requirements of the contract which must be changed if the proposal is adopted, and a recommendation as to how to make each such change (e.g., a suggested revision);

    (iii) an estimate of the reduction in performance costs, if any, that will result from adoption of the proposal, taking into account the costs of development and implementation by the Contractor (including any amount attributable to subcontracts in accordance with paragraph (e) below) and the basis for the estimate;

    (iv) a prediction of any effects the proposed change would have on collateral costs to the Government such as Government-furnished property costs, costs of related items, and costs of maintenance and operation;

    (v) a statement of the time by which a change order adopting the proposal must be issued so as to obtain the maximum cost reduction during the remainder of this contract, noting any effect on the contract completion time or delivery schedule; and

    (vi) the dates of any previous submissions of the proposal, the numbers of the Government contracts under which submitted, and the previous actions by the Government, if known.

(c) (1) Cost reduction proposals shall be submitted to the Procuring Contracting Officer (PCO). When the contract is administered by other than the procuring activity, a copy of the proposal shall also be submitted to the Administrative Contracting Officer (ACO). Cost reduction proposals shall be processed expeditiously; however, the Government shall not be liable for any delay in acting upon any proposal submitted pursuant to this clause. The Contractor does have the right to withdraw, in whole or in part, any value engineering change proposal not accepted by the Government within the period specified in the proposal. The decision of the Contracting Officer as to the acceptance of any such proposal under this contract (including the decision as to which clause is applicable to the proposal if this contract contains both a "Value Engineering Incentive" and a "Value Engineering Program Requirement" clause) shall be final and shall not be subject to the "Disputes" clause of this contract.

(2) The Contracting Officer may accept, in whole or in part, either before or within a reasonable time after performance has been completed under this contract, any cost reduction proposal submitted pursuant to this clause by giving the Contractor written notice thereof reciting acceptance under this clause. Where performance under this contract has not yet been completed, this written notice may be given by issuance of a change order to this contract. Unless and until a change order applies a value engineering change proposal to this contract, the Contractor shall remain obligated to perform in accordance with the terms of the existing contract. If a proposal is accepted after performance under this contract has been completed, the adjustment required shall be effected by contract modification in accordance with this clause.

(3) If a cost reduction proposal submitted pursuant to this clause is accepted by the Government, the Contractor is entitled to share in instant contract savings, collateral savings, and future acquisition savings not as alternatives, but rather to the full extent provided for in this clause.

(4) Contract modification made as a result of this clause will state that they are made pursuant to it.

(d) If a cost reduction proposal submitted pursuant to this clause is accepted and applied to this contract, an equitable adjustment in the contract price and in any other affected provisions of this contract shall be made in accordance with this clause and the "Termination for Convenience," "Changes," or other applicable clause of this contract. The equitable adjustment shall be established by determining the effect of the proposal on the Contractor's cost of performance, taking into account the Contractor's cost of developing the proposal, insofar as such is properly a direct charge not otherwise reimbursed under this contract, and the Contractor's cost of implementing the change (including any amount attributable to subcontracts in accordance with paragraph (e) below). When the cost of performance of this contract is decreased as a result of the change, the contract price shall be reduced by the following amount: the total estimated decrease in the Contractor's cost of performance less * percent (* %) of the difference between the amount of such total estimated decrease and any net increase in ascertainable collateral costs to the Government which must reasonably be incurred as a result of application of the cost reduction proposal to this contract. When the cost of performance of this contract is increased as a result of the change, the equitable adjustment increasing the contract price shall be in accordance with the "Changes" clause rather than under this clause, but the resulting contract modification shall state that it is made pursuant to this clause. (1967 JUN)

(e) The Contractor will use his best efforts to include appropriate value engineering arrangements in any subcontract which, in the judgment of the Contractor, is of such a size and nature as to offer reasonable likelihood of value engineering cost reductions. For the purpose of computing any equitable adjustment in the contract price under paragraph (d) above, the Contractor's cost of development and implementation of a cost reduction proposal which is accepted under this contract shall be deemed to include any development and implementation costs of a subcontractor and any value engineering incentive payments to a subcontractor, or cost reduction shares accruing to a subcontractor, which clearly pertain to such proposal and which are incurred, paid, or accrued in the performance of a subcontract under this contract.

(f) Omitted pursuant to ASPR 7-104.44(f).

(g) (1) A cost reduction proposal identical to one submitted under any other contract with the Contractor or another contractor may also be submitted under this contract.

(2) If the Contractor submits under this clause a proposal which is identical to one previously received by the Contracting Officer under a different contract with the Contractor or another contractor for substantially the same items and both proposals are accepted by the Government, the Contractor shall share instant contract savings realized under this contract, pursuant to paragraph (d) of this clause, but he shall not share collateral savings or future acquisition savings pursuant to paragraphs (f) and (j) (if included) of this clause.

(h) The Contractor may restrict the Government's right to use any sheet of a value engineering proposal or of the supporting data, submitted pursuant to this clause, in accordance with the terms of the following legend if it is marked on such sheet:

This data furnished pursuant to the Value Engineering clause of contract   shall not be disclosed outside the Government, or duplicated, used, or disclosed, in whole or in part, for any purpose other than to evaluate a value engineering proposal submitted under said clause. This restriction does not limit the Government's right to use information contained in this data if it is or has been obtained, or is otherwise available, from the Contractor or from another source, without limitations. If such a proposal is accepted by the Government under said contract after the use of this data in such an evaluation, the Government shall have the right to duplicate, use, and disclose any data reasonably necessary to the full utilization of such proposal as accepted, in any manner and for any purpose whatsoever, and have others so do.

In the event of acceptance of a value engineering proposal, the Contractor hereby grants to the Government all rights to use, duplicate or disclose, in whole or in part, in any manner and for any purpose whatsoever, and to have or permit others to do so, any data reasonably necessary to fully utilize such proposal.

(i) (1) For purposes of sharing under paragraph (d) above, the term "instant contract" shall not include any supplemental agreements to or other modifications of the instant contract, executed subsequent to acceptance of the particular value engineering change proposal, by which the Government increases the quantity of any item or adds any item, nor shall it include any extension of the instant contract through exercise of an option (if any) provided under this contract after acceptance of the proposal. Such supplemental agreements, modifications, and extensions shall be considered "future contracts" within paragraph (j) (if included) of this clause.

(2) If this contract is an estimated requirements or other indefinite quantity type contract, the term "instant contract" for purposes of sharing under paragraph (d) above shall include only those orders actually placed by the Government up to the time the particular value engineering change proposal is accepted. All orders placed subsequent to the acceptance of the particular change proposal shall be considered "future contracts" within paragraph (j) (if included) of this clause.

(3) If this clause is included in a basic ordering agreement, the "instant contract" for purposes of sharing under paragraph (d) above shall be the order under which the particular value engineering change proposal is submitted. Other orders under the same agreement shall be considered either "existing contracts" (if awarded prior to acceptance of the proposal) or "future contracts" (if awarded after acceptance of the proposal), within paragraph (j) (if included) of this clause.

(4) If this contract is a multi-year contract, the "instant contract" shall be the entire contract for the total multi-year quantity.

(j) Omitted pursuant to ASPR 1-1703.3(a)(1).   (ASPR 7-104.44(a), (c), (e)(i) and (f))

---

*Fifty percent (50%) for the first two approved proposals, fifty-five percent (55%) for the next two approved proposals, and sixty percent (60%) for all other approved proposals.

VALUE ENGINEERING INCENTIVE CLAUSE

1. *INTENT AND OBJECTIVES*—This clause applies to any cost reduction proposal (hereinafter referred to as a Value Engineering Change Proposal or VECP) initiated and developed by the Contractor for the purpose of changing any requirement of this contract. This clause does not, however, apply to any such proposal unless it is identified by the Contractor, at the time of its submission to the Government, as a proposal submitted pursuant to this clause.

1.1 VECPs contemplated are those that would result in net savings to the Government by providing either: (1) a decrease in the cost of performance of this contract, or; (2) a reduction in the cost of ownership (hereinafter referred to as collateral costs) of the work provided by this contract, regardless of acquisition costs. VECPs must result in savings without impairing any required functions and characteristics such as service life, reliability, economy of operation, ease of maintenance, standardized features, esthetics, fire protection features and safety features presently required by this contract. However, nothing herein precludes the submittal of VECPs where the Contractor considers that the required functions and characteristics could be combined, reduced or eliminated as being nonessential or excessive for the function served by the work involved.

1.2 A VECP identical to one submitted under any other contract with the Contractor or another Contractor may also be submitted under this contract.

2. *SUBCONTRACTOR INCLUSION*—The Contractor shall include the provisions of this clause, with the predetermined sharing arrangements contained herein, in all first-tier subcontracts in excess of $25,000 and any other subcontracts which, in the judgment of the Contractor is of such nature as to offer reasonable likelihood of value engineering cost reductions. At the option of the first-tier Subcontractor, this clause may be included in lower tier subcontracts. The Contractor shall encourage submission of VECPs from Subcontractors; however, it is not mandatory that VECPs be submitted nor is it mandatory that the Contractor accept and/or transmit to the Government VECPs proposed by his Subcontractors.

3. *DATA REQUIREMENTS*—As a minimum, the following information shall be submitted by the Contractor with each VECP:

3.1 A description of the difference between the existing contract requirement and the proposed change, and the comparative advantages and disadvantages of each; including justification where function or characteristic of a work item is being reduced;

3.2 Separate detailed cost estimates for both the existing contract requirement and the proposed change, and an estimate of the change in contract price including consideration of the costs of development and implementation of the VECP and the sharing arrangement set forth in this clause;

3.3 An estimate of the effects the VECP would have on collateral costs to the Government, including an estimate of the sharing that the Contractor requests be paid by the Government upon approval of the VECP;

3.4 Architectural, engineering or other analysis in sufficient detail to identify and describe each requirement of the contract which must be changed if the VECP is accepted, with recommendation as to how to accomplish each such change and its effect on unchanged work;

3.5 A statement of the time by which approval of the VECP must be issued by the Government to obtain the maximum cost reduction during the remainder of this contract, noting any effect on the contract completion time or delivery schedule; and,

3.6 Identification of any previous submission of the VECP including the dates submitted, the agencies involved, the numbers of the Government contracts involved, and the previous actions by the Government, if known.

4. *PROCESSING PROCEDURES*—Six copies of each VECP shall be submitted to the Contracting Officer, or his duly authorized representative. VECPs will be processed expeditiously; however, the Government will not be liable for any delay in acting upon a VECP submitted pursuant to this clause. The Contractor may withdraw, in whole or in part, a VECP not accepted by the Government within the period specified in the VECP. The Government shall not be liable for VECP development cost in the case where a VECP is rejected or withdrawn. The decision of the Contracting Officer as to the acceptance of a VECP under this contract shall be final and shall not be subject to the "Disputes" clause of this contract.

4.1 The Contracting Officer may modify a VECP, with the concurrence of the Contractor, to make it acceptable. If any modification increases or decreases the savings resulting from the VECP, the Contractor's fair share will be determined on the basis of the VECP as modified.

4.2 The Contracting Officer may accept, in whole or in part, a VECP submitted pursuant to this clause by giving the Contractor written notice thereof reciting acceptance under this clause. However, pending issuance of a formal change order or unless otherwise directed, the Contractor shall remain obligated to perform in accordance with the terms of the existing contract.

4.3 An approved VECP shall be finalized through an equitable adjustment in the contract price and time of performance by the issuance of a change order pursuant to the provisions of this clause bearing a notation so stating. Where an approved VECP also involves any other applicable clause of this contract such as "Termination for Convenience of the Government," "Suspension of Work," Changes," or "Differing Site Conditions" then that clause shall be cited in addition to this clause. **GSA FORM 2653 (8-71)**

5. COMPUTATIONS FOR CHANGE IN CONTRACT COST OF PERFORMANCE—Separate estimates shall be prepared for both the existing contract requirement and the proposed change. Each estimate shall consist of an itemized breakdown of all costs of the Contractor and all Subcontractors' work in sufficient detail to show unit quantities and costs of labor, material, and equipment.

5.1 Contractor development and implementation costs for the VECP shall be included in the estimate for the proposed change. However, these costs will not be allowable if they are otherwise reimbursable as a direct charge under this contract.

5.2 Government costs of processing or implementation of a VECP shall not be included in the estimate.

5.3 If the difference in the estimates indicate a net reduction in contract price, no allowance will be made for overhead, profit and bond. The resultant net reduction in contract cost of performance shall be shared as stipulated hereinafter.

5.4 If the difference in the estimates indicate a net increase in contract price, the contract price shall be adjusted pursuant to Clause 23 of the General Conditions of this contract.

6. COMPUTATIONS FOR COLLATERAL COSTS—Separate estimates shall be prepared for collateral costs of both the existing contract requirement and the proposed change. Each estimate shall consist of an itemized breakdown of all costs and the basis for the data used in the estimate. Cost benefits to the Government include, but are not limited to: reduced costs of operation, maintenance or repair, extended useful service life, increases in useable floor space, and reduction in the requirements for Government furnished property. Increased collateral costs include the converse of such factors. Computations shall be as follows:

6.1 Costs shall be calculated over a 20-year period on a uniform basis for each estimate and shall include Government costs of processing or implementing the VECP.

6.2 If the difference in the estimates as approved by the Government indicate a savings, the Contractor shall divide the resultant amount by 20 to arrive at the average annual net collateral savings. The resultant savings shall be shared as stipulated hereinafter.

6.3 In the event that agreement cannot be reached on the amount of estimated collateral costs, the Contracting Officer shall determine the amount. His decision is final and is not subject to the provisions of the "Disputes" clause of this contract.

7. SHARING ARRANGEMENTS—If a VECP is accepted by the Government, the Contractor is entitled to share in instant contract savings and collateral savings not as alternatives, but rather to the full extent provided for in this clause.

For the purposes of sharing under this clause, the term "instant contract" shall not include any changes to or other modifications of this contract, executed subsequent to acceptance of the particular VECP, by which the Government increases the quantity of any item of work or adds any item of work. It shall, however, include any extension of the instant contract through exercise of an option (if any) provided under this contract after acceptance of the VECP.

7.1 When only the prime Contractor is involved, he shall receive 50% and the Government 50% of the net reduction in the cost of performance of this contract.

7.2 When a first-tier Subcontractor is involved, he shall receive 30%, the prime Contractor 30% and the Government 40% of the net reduction in the cost of performance of this contract. Other Subcontractors shall receive a portion of the first-tier Subcontractor savings in accordance with the terms of their contract with the first-tier Subcontractor.

7.3 When collateral savings occur the Contractor shall receive 20% of the average one years net collateral savings.

7.3 The Contractor shall not receive instant savings or collateral savings shares on optional work listed in this contract, until the Government exercises its option to obtain that work.

8. DATA RESTRICTION RIGHTS—The Contractor may restrict the Government's right to use any sheet of a VECP or of the supporting data, submitted pursuant to this clause, in accordance with the terms of the following legend if it is marked on each such sheet:

The data furnished pursuant to the Value Engineering Incentive Clause of contract _____ shall not be disclosed outside the Government, or duplicated, used, or disclosed in whole or in part, for any purpose other than to evaluate a VECP submitted under said clause. This restriction does not limit the Government's right to use information contained in this data if it is or has been obtained, or is otherwise available, from the Contractor or from another source, without limitations. If such a proposal is accepted by the Government under said contract after the use of this data in such an evaluation, the Government shall have the right to duplicate, use, and disclose any data reasonably necessary to the full utilization of such proposal as accepted, in any manner and for any purpose whatsoever, and have others so do.

In the event of acceptance of a VECP, the Contractor hereby grants to the Government all rights to use, duplicate or disclose, in whole or in part, in any manner and for any purpose whatsoever, and to have or to permit others to do so, data reasonably necessary to fully utilize such proposal on this and any other Government contract.

GSA Form 2653 (8-71)

Example 1. Exterior Wall Sections*

GENERAL INFORMATION

PROPOSAL NO. __1__   DATE __1/27/71__

TITLE __EXTERIOR WALL SECTION-TEAM NO. 1__

Summary of Change (Description)

Before $ __25,420__   After $ __19,171__

Life Cycle Savings $ __12,010__

The present design for the wall section is 4-in. brick veneer on 8-in. block backup with a conventional parapet wall. Steel bar joists, with a metal deck, insulation, and a built-up roof, frame into this wall section. The foundation is block resting on a poured concrete footing.

Upon analyzing the total design, the team concluded that by adding additional structural steel columns to the north and south walls, the new frame system could permit a more efficient exterior nonbearing skin. In addition, the erection time would be less, thereby decreasing the total construction time. Also, this type of framing system better accommodates expansion than the wall-bearing system.

The limestone trim and the parapet were also eliminated.

The increased thermal efficiency of the skin in combination with that of the glass permitted the deletion of a hot water heating system by another team.

Estimated Cost Summary (see attached cost estimate)

	No. Of Units	Unit Cost	Total
A. Original			$25,420
B. Proposed			19,171
C. Contract savings (direct)			6,249
D. Other savings Maintenance			+
Operations Earlier occupancy: 2 mo @$5.00/sq ft			+
Miscellaneous −1/6 x 5.00 x 5200 sq ft			4,334
E. Gross savings (includes 35% OH + profit on contract saving only)			8,536
F. Implementing costs (if applicable)			860
Life-cycle cost savings			12,010

Percent savings __(25% contract)__

*Refer to Figure 42 for cost model of office building from which this wall was isolated as a high-cost, low-value component.

Exterior wall sections: Functional analysis worksheet.

ITEM Exterior Wall Section
(1) Control Elements
(2) Support Roof Load

QUAN	UNIT	COMPONENT	VERB	NOUN	KIND	EXPLANATION	VALUE OF WORTH	ORIGINAL COST
See	Lump sum	Limestone coping & Sill	Protect	Parapet & Wall	S			1)$5400
A&E		belt course	Shed	Water	S			1) 340
Est.		Bond beam R.C.	Support	Load	B	Lateral load dist.	360 (2)	2)1080
		Masonry backup	Support	Load	B	Vertical load dist.	3000 (2)	2)6140
		Foundation wall	Support	Load	B	Common item		1267
		Footing	Support	Load	B	Common item		1390
		Masonry facing	Control	Elements	B		2600 (1)	1)6000
			Provide	Facing	S			
			Provide	Aesthetics	S			
		Ceramic tile & glazing glass	Control	Elements	S			1) 640
			Provide	View	S			1)1920
		Asbestos panel & ceramic tile	Control	Elements	B		500 (1)	1)1200
		Metal frames	Hold	Sash & Spandrel	S			1)1600
		Flashing	Provide	Connection	S			1) 500
		Masonry reinforcement	Hold	Block	S			2) 500

(1) Worth of function Control elements: cost/worth = 17,6000/3,100 = 5.7
(2) Worth of function Support Roof Load: cost/worth = 7720/3360 = 2.3

PROJECT _____ OFFICE BUILDING _____

ITEM_____ Exterior Wall _____ PROJECT NO._____

BASIC FUNCTION _(1) Control Elements   (2) Support Load _____

DESIGN CRITERIA:

      (Mandatory to achieve the functions)

      20lb/sq ft live load -- roof loading

      Temp. +95° to +10°

      Footings are OK at 1500 lb/sq ft soil pressure

DESIGN HISTORY

    (Design criteria)

    No "U" factor specified

BACKGROUND

    Building located in small town

    TEAM MEMBERS

        Mr. H. Ray   Mr. J. Marion

        Mr. W. Bruce   Mr. J. Russo

        Mr. H. Gray

Exterior wall sections: Creative idea listing.

PROJECT _____ OFFICE BUILDING _____

ITEM _____ Exterior Wall _____ PROJECT NO. _____

BASIC FUNCTION _____ (1) Control Elements   (2) Support Load _____

UNINHIBITED CREATIVITY   DATE 1/27/71

(Don't Evaluate Idea . . . . . Idea Refinement is Later)

Leave Structure Alone

1. Should allow for future vertical expansion
2. Use brick in lieu of block and brick
3. Eliminate parapet & coping

4. Eliminate stone trim & sill-- use brick
5. ASTM C-270 Mortar

6. Make a symmetrical plan

7. Fenestration should be one piece
8. Piers should be brick size

9. Eliminate bond beam

Wall

10. Consider 8-in. brick wall

11. Consider 6-in. brick wall w/ styro & 5/8 gypsum finish
12. JM. Panel--Asbestos/Cement Insulation: Finish 1 side
13. JM. Panel--Asbestos/Cement Insulation: Finish 2 sides
14. US. Plywood Asbestos/Cement Insulation Panel: 1 side

15. US. Plywood Asbestos/Cement Insulation Panel: 2 sides
16. Spandrelite Panel Insulation

Change Structure to Frame

17. Add steel columns to the north and south walls
18. U.S. Plywood "Facade" full Height Panels Asb./Cement
19. Texture 1-11 wood panels on studs painted
20. Glasweld, insulated panel

21. J.M."Corespan" full height panels finish 2 sides
22.

23.

24.

25.

26.

27.

IDEAS	WEIGHTS								
CONSTRAINTS	Cost	Maintainability	Aesthetics	Proved Quality	Practical	Redesign Time	Score	Rank	
	4	3	3	3	1	2			
More steel cols.	4	2	3	4	4	2	51	3	
Insulated wall panels	3	4	4	4	4	1	54	1	
Texture 1-11	4	2	3	2	1	1	40		
Plywood	4	2	3	2	1	1	40		
Asb/cem panel	3	3	2	2	2	1	37		
8-in.brick wall	3	2	3	3	1	1	39		
Symmetrical plan	2	3	4	0	3	1	34		
No stone trim	4	4	4	0	4	4	52	2	
No parapet	4	4	4	0	4	3	50	4	
8-in. block wall	3	1	1	3	2	1	31		
No bond beam	3	0	0	0	1	1	16		
EXCELLENT 4									
GOOD 3 FAIR 2									
POOR 1 NONE 0									

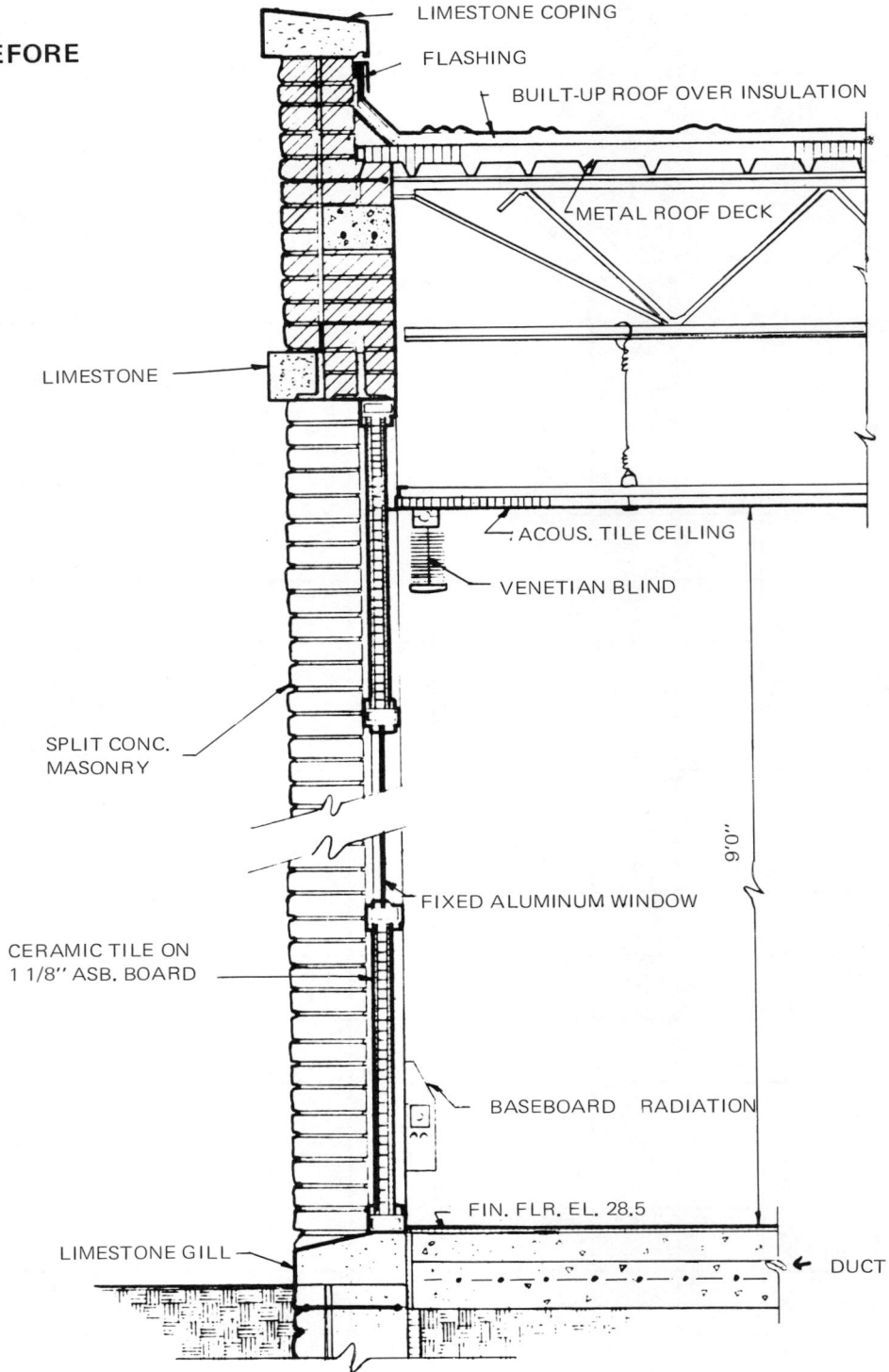

Exterior wall sections: Original plan.

BEFORE

LIMESTONE COPING

FLASHING

BUILT-UP ROOF OVER INSULATION

METAL ROOF DECK

LIMESTONE

ACOUS. TILE CEILING

VENETIAN BLIND

SPLIT CONC.
MASONRY

9'0"

FIXED ALUMINUM WINDOW

CERAMIC TILE ON
1 1/8" ASB. BOARD

BASEBOARD   RADIATION

FIN. FLR. EL. 28.5

LIMESTONE GILL

DUCT

Exterior wall sections: VE-revised plan.

Cost summary:

	Before	$25,420
	After	19,171
	Savings	$ 6,249

AFTER

FLASHING

BUILT-UP ROOF OVER INSULATION

COPING

METAL ROOF DECK

PRE-FABRICATED
INSULATED PANEL

ACOUSTICAL TILE CEILING

STEEL COLUMN

STEEL GIRT

INSULATED GLASS

9'0"

STEEL GIRT

COSTS

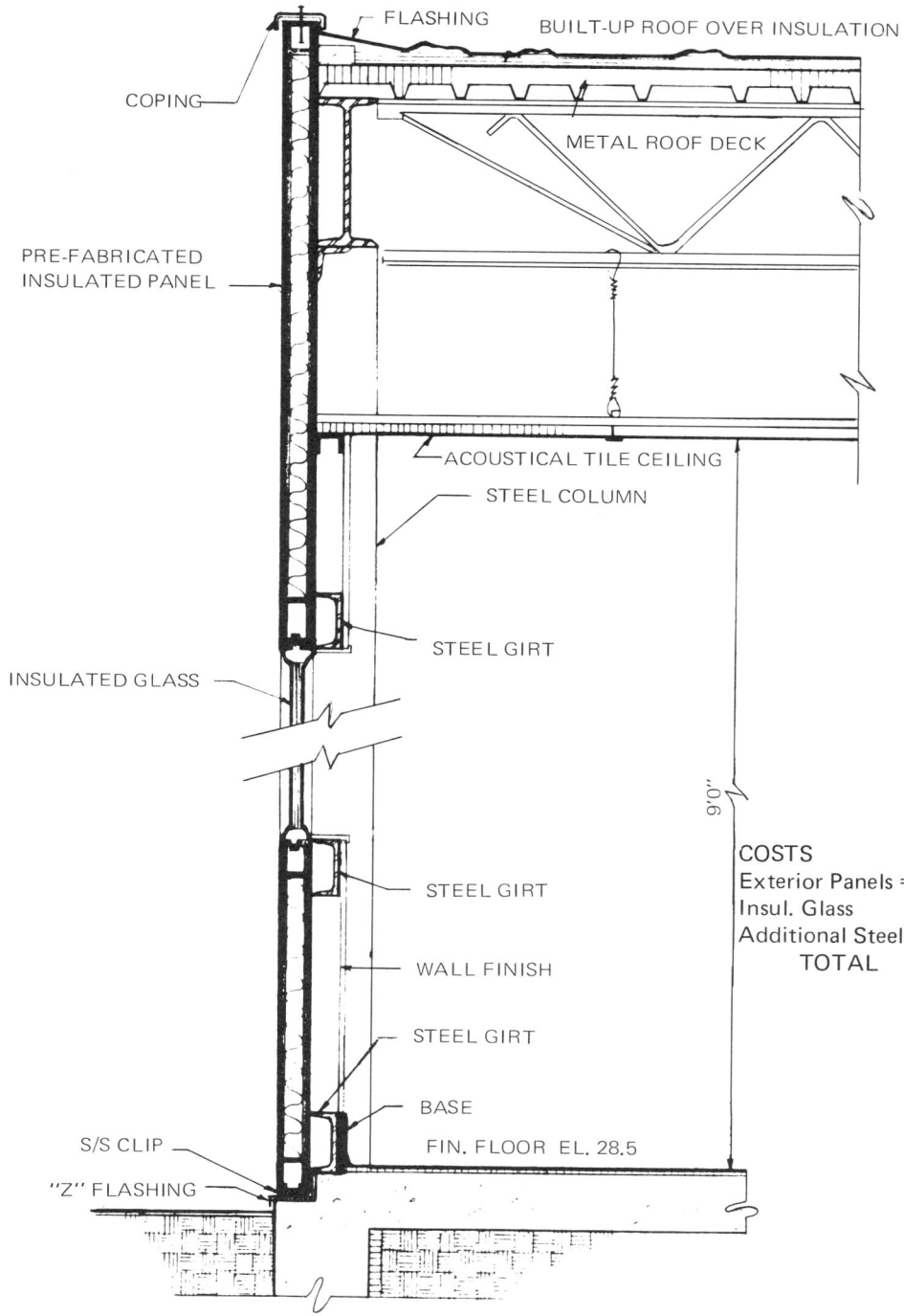

Exterior Panels =	$17,280
Insul. Glass	1,481
Additional Steel	410
TOTAL	$19,171

WALL FINISH

STEEL GIRT

BASE

S/S CLIP

FIN. FLOOR EL. 28.5

"Z" FLASHING

SCALE: 3/4" = 1'0"

Example 2. Reinforced Concrete Culverts

PROJECT ___RURAL INTERSTATE HIGHWAY_____

PROPOSAL NO. _S-1_____ DATE ___23 FEBRUARY 1971_____

    ITEM ___REINFORCED CONCRETE CULVERTS_____

                          POTENTIAL SAVINGS _____

                                   INSTANT ___$220,000_____

                                LIFE CYCLE ___N.A._____

I.   INFORMATION PHASE

    A.   GENERAL

        1.   Design Criteria

            AASHO Standard Specifications for Highway Bridges, 1965 Edition, Modified

            Concrete: Class A, $f'c = 3000$ lb/sq in, $fc = 1200$ lb/sq in

            Reinforcement: ASTM A15 with deformations according to ASTM A305, $fs = 20000$ lb/sq in, minimum lap 24 diameters

            Live load: HS 20-44

        2.   Background

            The culvert is a rectangular twin box of reinforced concrete.

    B.   FUNCTIONAL ANALYSIS

            See attached worksheet for functional analysis.

            The worth of the basic function was calculated using corrugated pipe and choosing the optimum size and number for the volume required.

Discussion

The total estimated cost of a typical reinforced concrete box culvert for a rural highway is approximately $400,000. This cost can be itemized as follows:

Concrete for culvert, approximately $300,000 or 75% of the cost.

Bar reinforcement, approximately $50,000 or 12.5% of the cost.

The other 12.5% covers other smaller items as shown in functional analysis worksheet.

The class A concrete, because of its high cost and a cost/worth ratio of 300/130=2.3 represents the area of greatest savings potential; therefore, ·the VE team decided to concentrate its effort in this area.

II. SPECULATIVE PHASE

A. IDEAS GENERATED

During the team's brainstorming session, the ideas shown in the creative idea listing were generated to accomplish the basic function "provide drainage."

III. ANALYTICAL PHASE

A. EVALUATION OF IDEAS

The team considered the ideas generated and decided that the second idea (reinforced concrete multiple-cell box culvert), the fifth (corrugated steel pipe), and the twelfth (precast, post-tensioned concrete sections), the greatest savings potential. The other ideas were dropped as being too costly, obsolete, or not practical.

B. INVESTIGATION OF ALTERNATES

The alternates generated were developed as three separate studies, A-1, A-2, and A-3, which will be discussed later in this report.

Study A-1 considers the existing configuration and concentrates on affecting economies through a reappraisal of the design.

Study A-2 considers a change in concept through the use of corrugated steel pipe.

Study A-3 considers a change in concept through use of precast post-tensioned concrete.

STUDY A-1:   EXISTING CONFIGURATION
   For immediate implementation, the team decided to conduct a critical reappraisal of the design criteria of the existing culvert with the objective of reducing concrete quantities. The team's review revealed that a unit weight of 120/cu ft for earth was specified and no reduction in live load permitted. The AASHO Specifications (Article 1.2.2 A), state that "for box culverts and culverts with cast-in-place inverts or footings, the weight of the earth may be taken at 70 percent of its actual weight." The same specifications (Article 1.3.3), allow that "for multiple spans it(the effect of live load) may be neglected when the depth of fill exceeds the distance between faces of end supports."
   To document the potential savings, a typical twin box culvert was used as an example. For design data and cost analysis see Engineering Design Data.*

STUDY A-2:   CORRUGATED STEEL PIPE
   The team decided to conduct a thorough investigation of using corrugated steel pipe or multiplate pipe as a sub-stitute for the reinforced concrete culvert.
   Price information received from various manufacturers indicates that both the corrugated steel pipe and the multiplate pipe schemes are more economical than the reinforced concrete culvert.
   The attached tables* show the in-place cost per linear foot for a 3 in. by 1 in. corrugated galvanized steel pipe and for a 6 in. by 2 in. multiplate round pipe.

STUDY A-3
   Study A-3 was initiated but suspended due to lack of
   *Not included here, but included in actual report.

time and funds. Preliminary investigation indicated substantial interest in this area by manufacturers of precast concrete sections, but questionable supply and inability to procure meaningful prices forced the team to suspend investigation.

IV. RECOMMENDATIONS

Study A-1

The team's study of the typical reinforced concrete culvert revealed that significant potential savings were possible in the concrete quantities and cost of structure. The team's functional analysis revealed that concrete and reinforcement were potential areas for savings. The attached sketch shows the existing detail and the alternate detail that were developed. For technical data and cost analysis, see Engineering Design Data.*

The team recommends use of the alternate detail for future use. Estimated savings for a typical reinforced concrete culvert are approximately $74,000, or 18% of the total cost (see sketch of original and VE-revised plans).

Study A-2

The team's study revealed significant potential savings by using three, 120-in. diameter corrugated galvanized steel pipe.
For technical data and cost analysis, see Engineering Design Data.*
The team recommends corrugated galvanized steel pipe for future use. Estimated savings for a typical project are approximately $222,627 or 54.7% of the total cost (see sketch or original and VE-revised plans).

   *Not included here.

Reinforced concrete culverts: Functional analysis worksheet.

ITEM REINFORCED CONCRETE CULVERTS

FUNCTION PROVIDE DRAINAGE

QTY	UNIT	COMPONENT	FUNCTION VERB	FUNCTION NOUN	FUNCTION KIND	EXPLANATION	VALUE OF WORTH	ORIGINAL COST
570	cu yd	Selected granular fill	Control	Grade	S			$24,990
			Support	Load	S			
			Protect	Culvert	S			
900	cu yd	Structure excavation	Provide	Access	S			6,426
700	cu yd	Class A concrete for structures	Provide	Drainage	B	216 sqft of area	$130,000	293,700
			Transmit	Load	S	Compressive stress		
			Protect	Reinforcement	S	Corrosion		
			Distribute	Load	S			
110	cu yd	Class B concrete for structures	Transmit	Load	S	Compressive stress		14,300
			Protect	Reinforcement	S	Corrosion		
			Distribute	Load	S			
			Control	Erosion	S			
86,000	lb	Bar reinforcement for structures	Transmit	Load	S	Tensile stress		57,200
			Prevent	Cracking	S			
			Distribute	Load	S			
		Miscellaneous			S			9,470
						TOTAL	$130,000	$406,086

Cost/Worth=406/130=3.1

PROJECT ___RURAL INTERSTATE HIGHWAY___

ITEM ___REINFORCED CONCRETE BOX CULVERT___ PROJECT NO.___S-1___

BASIC FUNCTION ___PROVIDE DRAINAGE___

(Don't Evaluate Idea . . . . . Idea Refinement is Later)

Uninhibited Creativity   DATE ___February 23, 1971___

1. Reinforced concrete single-cell box culvert
2. Reinforced concrete multiple-cell box culvert
3. Reinforced concrete single-cell bridge culvert
4. Reinforced concrete multiple-cell bridge culvert
5. Corrugated steel pipe
6. Corrugated steel pipe arch
7. Plain concrete pipe
8. Reinforced concrete pipe
9. Reinforced concrete arch
10. Corrugated metal on reinforced concrete foundation
11. Stone masonry arch on reinforced concrete foundation
12. Precast, post-tensioned concrete sections
13. Precast reinforced concrete sections
14. Ditch
15. Tunnel
16.
17.

18.
19.
20.
21.
22.
23.
24.
25.
26.
27.
28.
29.
30.
31.
32.
33.
34.

Reinforced concrete culverts: Original plan and VE-revised plan for using existing
    configuration with less concrete.
Cost summary: Before $406,000
                 After   332,000
                 Savings $ 74,000 (18%)

BEFORE

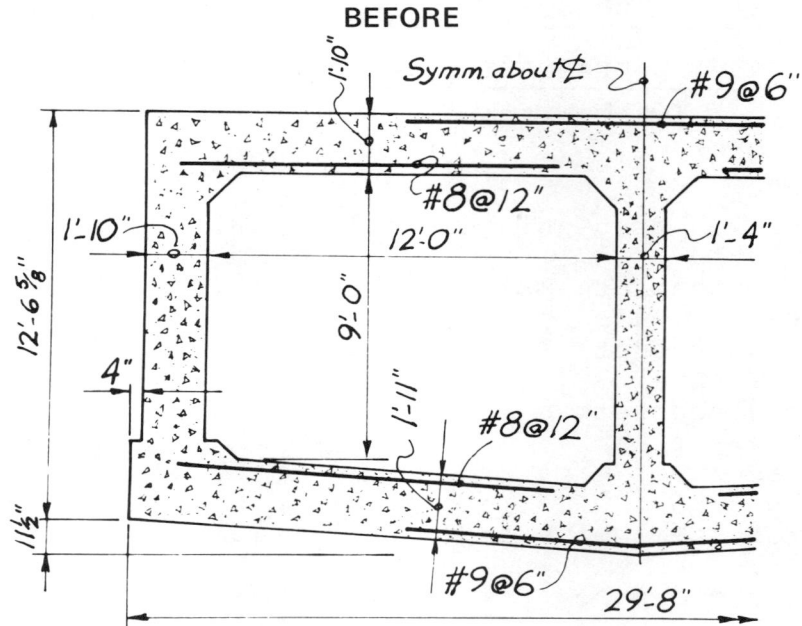

AFTER

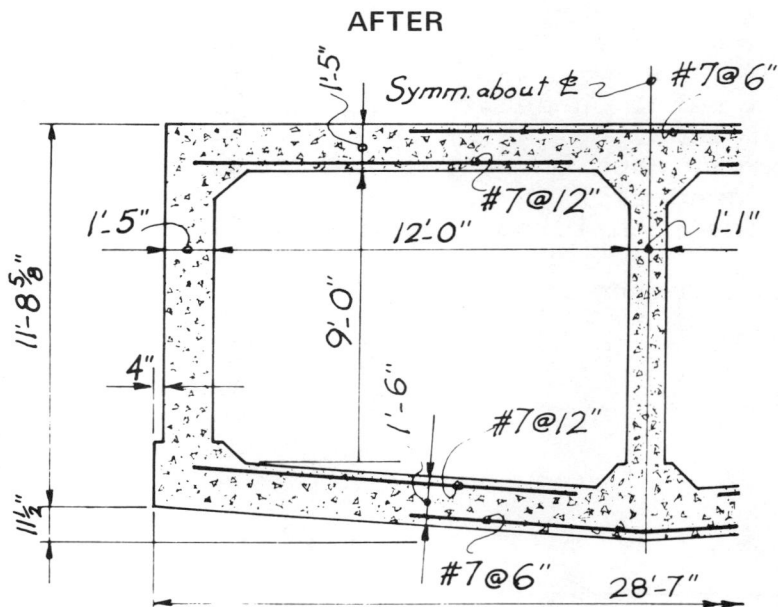

Reinforced concrete culverts: Original plan and VE proposal for replacement with corrugated galvanized steel pipe.

Cost summary: Before $406,000
                    After   186,000
                    Savings $220,000 (54%)

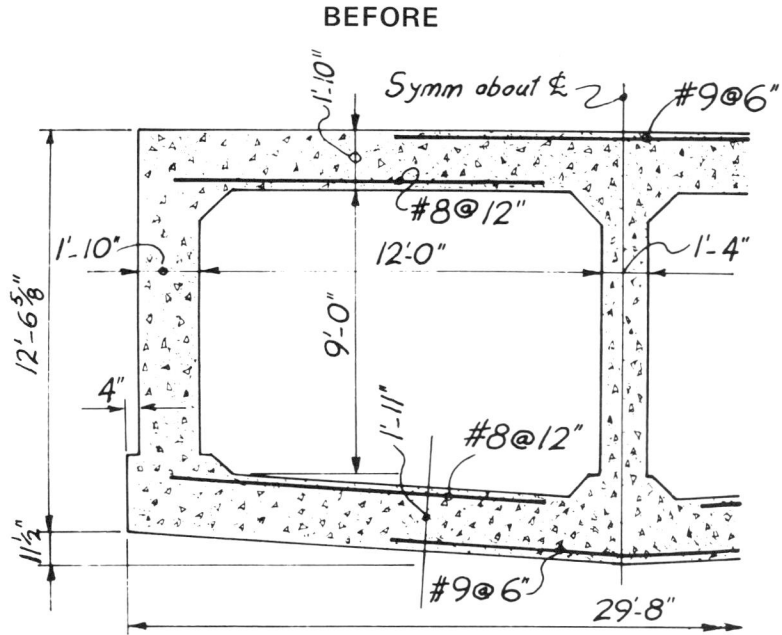

BEFORE

Symm about ℄

#9@6"

1'-10"

#8@12"

1'-10"   12'-0"   1'-4"

12'-6⅝"

9'-0"

4"

1'-11"

#8@12"

1½"

#9@6"

29'-8"

AFTER

120"

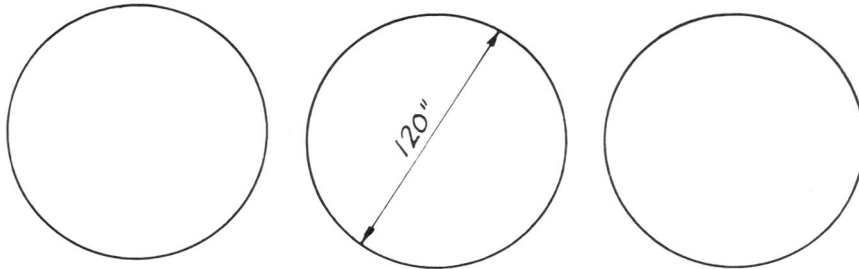

3-120"⌀, 3"x1", 10Ga Corrugated Galvanized Steel Pipes

APPENDIX C. THE ETHICS OF VE

What if your boss asks you to review his pet civic project plans & specs?

This can be sticky. You have to consider what he wants from you in the way of actual engineering review. Then there is the problem of advising the architect involved of any engineering review of the project.

Case No. 68-11

Subject

Review of Architect's plans—Section 2 (b)—Code of Ethics; Section 6—Code of Ethics; Section 12 (a) —Code of Ethics; Section 12 (b)— Code of Ethics; Section 15 (a)— Code of Ethics.

Facts

Engineer "A" is employed by an industrial corporation. His immediate supervisor is Engineer "B", who is chairman of a civic committee responsible for retaining an architect to design a civic facility. When Engineer "B" received the completed plans and specifications from the architect, he directed Engineer "A" to review them in order to (1) gain knowledge, (2) suggest improvements, (3) assure their compliance with the specified requirements.

Questions

1. Is the instruction of Engineer "B" to Engineer "A" consistent with the Code of Ethics?
2. Is Engineer "A" ethically permitted to carry out the instructions given him by Engineer "B"?

References

Code of Ethics—Section 2 (b)— "He shall seek opportunities to be of constructive service in civic affairs and work for the advancement of the safety, health, and well-being of his community." Section 6—"The Engineer will undertake engineering assignments for which he will be responsible only when qualified by training or experience; and he will engage, or advise engaging, experts, and specialists whenever the client's or employer's interests are best served by such service." Section 12 (a)—"An Engineer in private practice will not review the work of another engineer for the same client, except with the knowledge of such engineer, or unless the connection of such engineer with the work has been terminated." Section 12 (b)—"An Engineer in governmental, industrial, or educational employ is entitled to review and evaluate the work of other engineers when so

required by his employment duties." Section 15 (a)—"He will encourage his engineering employees' efforts to improve their education."

Discussion

We base our discussion and conclusions on the assumption that Engineer "B" as a supervisor was acting within the scope of his authority and had the explicit or implied permission of his employer to use his time and that of Engineer "A" for the stated purposes.

Both Engineer "A" and Engineer "B" would be providing professional services in accord with the mandate of Section 2 (b) in that their respective roles are intended to be of constructive value to the community.

Also, Engineer "B" is acting within the concept of Section 15 (a) in having Engineer "A" perform the review for the purpose, in part, of gaining knowledge.

The more pertinent and difficult question, however, is the limitation imposed by Section 12 (b). This section of the Code implies that the review of the work of another engineer by an engineer in industry is for the purposes of his employer. Proceeding on the assumption that the review is being done with the consent of the employer, we can rationalize that a civic improvement project is for the benefit of the employer, which is a part of the community.

We turn then to the interprofessional aspects of the question. Here the review is of the work of an architect, not of another engineer. Section 12 (a) does not provide any specific guidance because it is restricted to engineers in private practice and deals only with work for the same client. Nevertheless, we believe it provides a general guideline in the sense that it expresses the principle that in that type of case the reviewing engineer has a duty to ascertain that the designing engineer has knowledge of the review unless the connection of the designing engineer with the client has been terminated. The last proviso is not applicable here.

Even though the Code refers only to engineers, we believe that the principles enunciated in these circumstances should be applicable to other professions, and particularly to other design professions. This view is supported by the document, "Professional Collaboration in Environmental Design," approved both by NSPE and the American Institute of Architects. It states, in part, that engineers and architects, ". . . perform their services in accordance with the standards of conduct and code of ethics of their individual professions, and each respects the standards and codes of the other profession."

In light of this principle, we believe it would be incumbent upon Engineer "B" to advise the architect of the proposed review of his plans and specifications by Engineer "A" in order that the architect may have the opportunity to comment upon such evaluation or recommendations as may be submitted by Engineer "A" to the civic committee through Engineer "B"; but in accordance with Section 6, Engineer "A" must limit his suggestions and opinions to those aspects within his competence.

Conclusions *

1. The instruction of Engineer "B" to Engineer "A" would be consistent with the Code of Ethics, provided that Engineer "B" advised the architect of the proposed review.

2. Engineer "A" is ethically permitted to carry out the instructions given him by Engineer "B" for those aspects within his competence.

*Note—This opinion is based on data submitted to the Board of Ethical Review and does not necessarily represent all of the pertinent facts when applied to a specific case. This opinion is for educational purposes only and should not be construed as expressing any opinion on the ethics of specific individuals. This opinion may be reprinted without further permission, provided that this statement is included before or after the text of the case.

Appendix D. VE Review (Expanded Idea Listing) for Senior High School (Approximate Total Estimate of School $3,5000,000)

Summary

Maximum Savings Potential—Before Owner and A&E Review

Areas	Number of Items	Potential Savings
Architectural	14	$139,000
Plumbing	8	14,000
Mechanical	7	59,500
Electrical	8	18,500
Total	37	$231,000

Savings Potential After Owner and Architect-Engineer Review

Based on the review held at the owner's office, savings of over $40,000 can be implemented immediately by the architect-engineer. An additional $60,000 savings potential requires further in-depth analysis using the VE job plan.

Note: Since this review was conducted on almost completed documents, its intent was to isolate areas of potential savings without extensive changes to existing design documents.

DESCRIPTION	REFERENCE	ESTIMATED POTENTIAL SAVINGS		REMARKS
		INITIAL	LIFE CYCLE	
ARCHITECTURAL A-1 Specs call for 3-coat floor hardener application. This seems excessive. Suggest use of one-coat system and opening requirements for other manufacturers. Note: Over 100,000 sq ft involved.	Spec 2-2-14 Para (b)	Up to $15,000	Question-able	Easily implemented. To be implemented.
A-2 Specs call for 20-year bonded roof. Suggest not exercising bond since government agency experience indicates little value for bond.	Spec 2-6-1&2-6-2	$2,000	N/A	Note: Contractor guarantee also required for 2 years. This should be adequate. Dropped.
A-3 Plans and specs call for a formed acrylic fascia board around building. If cost is important factor, consider use of a more standard section such as Galbestos or Butler panel.		Up to $25,000		A trade-off in aesthetics. Easily implemented. Included as an alternate in Amendment # 3.
A-4 Plans and specs call for some areas to be 1-in. plaster for ceilings & firewall, and 3/4-in. thick for others. Consider substituting (optional) 5/8-in. fire-rated gypsum board and thin coat plaster for 1-in. plaster and thin coat for other areas.		Up to $10,000	Some sacrifice	Minimum redesign Will be reviewed.

DESCRIPTION	REFERENCE	ESTIMATED POTENTIAL SAVINGS		REMARKS
		INITIAL	LIFE CYCLE	
A-5 Specs and Amendment #2 call for 1-in. spray-in-place rigid urethane foam on exterior face of block. Suggest option to allow contractor to use 1-in. rigid urethane board (inside face) as an alternate. If weather conditions are adverse, spray-in-place cannot be applied while rigid board can be utilized.	Amend. No. 2 page 4 2.4-4	Up to $3,000	N/A	Minimum redesign. Note: By requiring foam on outside, natural construction sequence of masonry is disturbed. To be studied in depth.
A-6 a. Specs call for four coats of paint for interior woodwork of hardwood. Suggest option to use factory finish wherever possible, and prefabricated assembly.	2-17-3 Para 17-13 (a)	Up to $5,000	N/A	Savings of approximately $20/door. Easily implemented. Owner to check with union.
b. Specs call for interior block to have two coats of flat latex. This will show dirt easily and is not too washable. Suggest use of semigloss such as TT-P-95 which requires less maintenance.	2-17-4 Para 17-16 (a)	N/A	Yes	Easily implemented. To be implemented.
A-7 Specs call for use of double-glazed window. If costs are critical for the area involved, and since there are ventilating lights in most sash which increase infiltration, consider use of single 7/32-in. Graylite glass.	A-8 2-16-1(b)	Up to $10,000	Some adverse impact	A trade-off item. Easily implemented. Note:Double glazing costs approx. $6.50 sq ft; single glazing $3.50 Electric co. require; dropped

DESCRIPTION	REFERENCE	ESTIMATED POTENTIAL SAVINGS		REMARKS
		INITIAL	LIFE CYCLE	
A-8 Specs and plans call for stock bond for interior masonry. This is expensive. Suggest use of regular bond.		Up to $4,000		Easily implemented. Dropped.
A-9 Specs call for contractor to furnish carpet. Suggest deletion from contract and consider direct purchase.	2-15-3 Para 15-10(a)	Up to $6500		Easily implemented. Owner to investigate; must be investigated for coordination between general contractor and carpet supplier
A-10 Specs call for rigid insulation on top of insulating concrete. Question value of rigid insulation in view of costs. Consider deletion if costs are critical.	Sections 5 and 6 2-6-2 Para 6-02(c)	Up to $35,000	Some adverse impact	A trade-off item. Easily implemented. Requires additional information; mechanical to provide. To be studied in depth.
A-11 Specs call for a 6-in. macadam base under pavement. For economy, consider substitution of a compacted, crusher-run base course.	2-3-3	Up to $10,000	N/A	Easily implemented. To be implemented. Architect-engineer to develop specs.

| DESCRIPTION | REFERENCE | ESTIMATED POTENTIAL SAVINGS | | REMARKS |
		INITIAL	LIFE CYCLE	

A-12 Finishes. If cost savings are critical, with minimum impact on needed requirements consider the following:
 	
   a. In lieu of cement enamel for special wall areas consider using all epoxy wall finishes.
 	
   b. In lieu of all areas specified for vinyl wall covering reduce area 50% and use epoxy wall finishes in deleted areas.
	2-17-4
	2-17-5
	Para
	17-21
	17-22
		Up to $3500
		
		Up to $10,000
			N/A	Easily implemented. Changed by Amendment to an alternate.
				
				Easily implemented. Minimum redesign required. To be considered if savings needed to realize project.

DESCRIPTION	REFERENCE	ESTIMATED POTENTIAL SAVINGS		REMARKS
		INITIAL	LIFE CYCLE	
PLUMBING				
P-1 Specs call for painting of all apparatus, equipment, etc.,in mechanical spaces. This requirement is costly and should be reviewed for minimum requirements and allowances to use factory-supplied coatings.	Spec 4-1-3(a)	Up to $1000	Minimum impact	Easily implemented. Note to be added to specs.
P-2 Specs call for use of 6lb lead for flashing sleeves. Common practice is use of 4-1b. Save approximately $3/drain.	Spec 4-1-4(b)	Up to $500	N/A	Easily implemented. Dropped.
P-3 a. Specs call for use of "M" type copper for DWV above grade.Why not allow use of DWV grade and save additional money?	4-3-2 Para 3-02(b)	Up to $2000	N/A	Easily implemented. To be implemented.
b. Specs call for use of extra heavy C I pipe above ground for 3-in. size and above. Is this a code requirement? Functionally, standard weight pipe would last longer than building.		Up to $5000		To be implemented.

DESCRIPTION	REFERENCE	ESTIMATED POTENTIAL SAVINGS		REMARKS
		INITIAL	LIFE CYCLE	
c. Specs call for standard galvanized pipe for sanitary vent piping. Suggest optional requirement of copper and cast iron for small sizes.	Varies			To be implemented.
P-4 Specs call for use of galvanized steel pipe for compressed air lines. Black steel should be adequate. Consider allowing its use.	4-3-3(g)	Up to $1000	N/A	Note: Gas piping not galvanized. To be implemented.
P-5 Specs call for Duriron acid waste and vent piping. Nalgene and GSR make plastic pipe for acid DWV which was proved satisfactory and costs substantially less than Duriron. Consider opening spec.	4-3-3(j)	Up to $4000	Questionable	Nalgene and GSR also supply tanks and accessories Fuseal joints by GSR are recommended. To be studied in depth.
P-6 Specs call for use of Speakman flush valves. Allow use of Sloan and Coyne-Delaney equal valves. Savings of $2-3/valve possible.	4-6-2 (a) & (b)	Up to $500	N/A	Easily implemented. To be implemented.

DESCRIPTION	REFERENCE	ESTIMATED POTENTIAL SAVINGS		REMARKS
		INITIAL	LIFE CYCLE	
MECHANICAL				
M-1 Plans call for use of XH cast iron pipe for enclosure of underfloor exhaust duct. Question value. Recommendation: Suggest use of asbestos cement pipe and deletion of XH cast iron pipe. Consider use of transit pipe for duct.	Plans 5-7-1 (a)	Up to $500	N/A	Requires some redesign. Dropped. Specified for reliability.
M-2 Plans and specs call for two large central domestic hot water heaters. Suggest use of decentralized system. This type of building very spread out and piping costs are high.		Up to $2000	Question-able	Requires some redesign and a possible compromise solution. Dropped.
M-3 Specs call for all air-handling units, motors, and pumps to be provided with three coats of paint applied at the factory. Suggest use of manufacturer's standard and some sort of a performance spec. Many satisfactory coatings are available in less than three coats.	Spec 5-1-5(c)	Up to $1000	N/A	Can be readily implemented. To be implemented.

DESCRIPTION	REFERENCE	ESTIMATED POTENTIAL SAVINGS		REMARKS
		INITIAL	LIFE CYCLE	
M-4 Specs and plans call for underground exhaust system. If savings desired, consider use of an overhead system. More flexibility and less costs will result.	Spec 5-7-1 Para 7-02	Up to $1000	Yes	Requires minimum redesign. To be reviewed.
M-5 Specs call for all air-conditioning system supply air duct not acoustically lined to have rigid insulation 1 in. thick. This type of insulation is expensive. Flexible blanket suitable for all areas exept where physical damage likely.	Spec 5-8-2 5-8-3 Para 8-03 (a) & 8-04	Up to $30,000	N/A	To be studied in depth.
M-6 Specs call for automatic pneumatic control system for HVAC system. If costs are critical consider use of an electric control system.	Spec 5-9-1 Para 9-01	Up to $5- 10,000	See remarks	Check page 5-11-1 Requires some redesign with some sacrifice in long-term costs. Dropped. Local condition dictate bids more favorable toward pneumatic.

| DESCRIPTION | REFERENCE | ESTIMATED POTENTIAL SAVINGS | | REMARKS |
		INITIAL	LIFE CYCLE	
M-7 Plans list ventilating air rates for various areas. Rates could be reduced to reduce costs and meet needed requirements.	H-1	Up to $15,000	N/A	Requires major redesign. Dictated by state requirements. To be studied in depth for future projects.

DESCRIPTION	REFERENCE	ESTIMATED POTENTIAL SAVINGS		REMARKS
		INITIAL	LIFE CYCLE	
ELECTRICAL E-1 Specs call for minimum conduit size to be 3/4 in. Suggest use of 1/2 in. conduit for branch circuits and single switch runs in accordance with NEC code. Equal capacity can be achieved through use of THWN or XHHW wire. Note: Above applies to conduit for signal and control circuits.	Spec 6-2-3 Para 2-04(b)	Up to $5000		Easily implemented. Classified on drawings. Allowed for signal and control.
E-2 a. Specs call for use of RHW wire in all first floor slabs, outside wall etc. Suggest use of less expensive wire that will meet requirements such as THWN or XHHW.	P-6-2-5 Para (b) P6-3-3 Para 3.08 (b)	Up to $2000		Save approximately 20% for material and possible reduction of conduit size. Dropped due to local experience.

DESCRIPTION	REFERENCE	ESTIMATED POTENTIAL SAVINGS		REMARKS
		INITIAL	LIFE CYCLE	
b. Specs call for exclusive use of copper bus duct. Suggest optional use of aluminum bus duct. Note: Stress use of proper connectors.	6-3-3 Para 3-08 (a)	Up to $1000		To be implemented.
E-3 Specs call for exclusive use of steel conduit. Suggest optional use of aluminum conduit where exposed. Lighter weight and less labor with resulting savings can be realized.	6-2-5 Para 209(a)	Up to $1000		Easily implemented. To be implemented.
E-4 Specs call for soldered splices and no wire nuts. This is an expensive require-ment. Suggest use of pressure connectors similar to Buchanan or Scotch's spring connectors which are presently being used extensively in commercial areas.	6-2-5 Para 2-10(f) 6-4-1 Para 4-01(f)	Up to $2,000		This appears to be an absolute method (Changed by Amendment No.3) No longer applicable.

DESCRIPTION	REFERENCE	ESTIMATED POTENTIAL SAVINGS		REMARKS
		INITIAL	LIFE CYCLE	
E-5 Specs call for two 4-in. steel conduits for electric service. Suggest consideration of asbestos cement or fiber duct.	6-3-2 Para 306 (a)	Up to $500		Easily implemented. Approximately 50% savings (Changed by Amendment No.3 to PVC.) No longer applicable.
E-6 Specs and plans call for main secondary breakers to be air circuit breakers type DB-50 by Westinghouse. For savings use of fused type breakers could be considered.	6-3-4 Para (a)	Up to $5000		Some redesign involved. Dropped. Required for ease of maintenance. To be studied in depth for future projects
E-7 Plans and specs call for significant use of two-tube fluorescent fixtures in areas where the possibility of use of four-tube fixtures exists. Consider greater use of four-tube fixtures. Note: In order to properly review the electrical design it would be necessary to have design calculations to observe use of diversity factors, load development, etc.	Plan E-Series	Up to $2000		Some trade-off in lighting coverage. NEC Code for Schools allows no diversity.

APPENDIX E. PROPOSAL FOR VE STUDY

INTRODUCTION

The Contractor will provide approximately   * manhours of services necessary to conduct a   ** week Value Engineering study of a facility selected by the Client. The purpose of the study will be to develop cost reduction proposals to effect immediate and lifecycle savings to the Client. Proposals of cost reduction resulting from this study will be incorporated into a final written report.

Phase I. Scope of Work (General)

The Contractor will provide a multidisciplinary team including architects, professional engineers, and specialists experienced in facility design and the application of Value Engineering techniques.

During the Value Engineering study, the Contractor will:

1. Confirm and validate cost estimates
2. Isolate significant high-cost areas
3. Develop alternate methods to reduce costs and/or increase efficiency
4. Validate his findings to the maximum extent possible
5. Present results
6. Prepare a final report that includes finished proposals for cost reduction

In addition to challenging the material and design selections of the designer, the Contractor's team will analyze Client standards and criteria, which effect high costs in the structure, and propose changes to effect economies.

Because this contract represents an initial effort to formally apply Value Engineering to a Client facility, close coordination will be required between the contractor and client.

At minimum, the project plans, specifications, cost estimate, design data sheets, and calculations will be required as Client-furnished information. If necessary, and agreed to by the Client, a field trip (outside the contract scope) will be taken to the facility site.

*Varies from 500 to 1500 manhours depending on project scope, complexity, and client involvement.

**Varies from 4 to 12 weeks depending on urgency of project start and magnitude of scope of work.

The Contractor's team will then review and evaluate the collected data and break down the facility to identify component functions in order to isolate specific high-cost areas. At this time, the total resources of the Contractor's firm will be used by gathering a team at the main office. The team will make a list of isolated areas where there are indications that costs may be reduced and/or efficiency increased.

This phase will be completed upon the submission of the idea listing.

Phase II. Development and Evaluation of Solutions

Based on a joint review of the idea listing, during a period not exceeding 2 weeks, high-cost areas will be selected for in-depth study, and the Client will authorize contractor Phase II efforts. The areas selected must fall within contract scope.

Using the team approach characteristics of Value Engineering methodology, the Contractor's team will generate ideas which can be used for the development of alternate solutions to the selected high-cost areas.

Subsequently, workable ideas will be refined, evaluated, and developed into practical alternate solutions with a cost analysis prepared for each solution. In each cost analysis, due consideration will also be given to operating and maintenance costs. At this time the entire resources of our firm will be utilized by generating a team effort at our main office.

It is during this phase that outside consultants will be utilized and field trips* taken as directed and approved.

The team will document the most promising alternates for each high-cost area considered. Selections will be based on all pertinent factors, e.g.: aesthetics, initial and operating costs, safety, and maintainability. Subsequently, the Contractor will develop drafts of value engineering proposals (VEPs) and conduct an oral presentation of tentative results to selected Client personnel.

Phase III. Preparation and Presentation of Results

The Client will evaluate the tentative VEPs and forward comments to the Contractor during a period not to exceed 2 weeks. The Contractor will revise the alternate proposals based on the comments generated during the oral presentation and as a result of the Client's review. The revisions will be made with consideration given to the impact of each proposal on the overall design. Implementation procedures will be developed for each proposal. Our proposal format

*To be direct reimbursable item.

will follow the basic procedures used for this report.

The principal items to be covered in the final proposal are:

1. Description of the areas studied and the results attained, including appropriate sketches
2. Cost analyses of the present and proposed designs including backup quotations and detailed cost breakdowns
3. Technical backup required to support selection of alternates

The contract will be considered complete upon delivery of 10 copies of the final report. Additional copies of the final report will be furnished as requested at cost.

APPENDIX F. POSITION DESCRIPTION: GENERAL ENGINEER (VALUE ENGINEER)

Incumbent is the focal point and authority for the VE program. VE is the organized effort of critically reviewing and analyzing the technical aspects of both the design and the construction of selected individual projects to provide the required facility at the lowest overall costs consistent with requirements for performance, reliability, and maintainability. The complete spectrum of engineering specialization are represented in the construction program with the civil, mechanical, and electrical engineering disciplines dominating. A further duty of the incumbent is to foster an increased cost-consciousness among engineering personnel and their field representatives.

Principal Duties and Responsibilities

1. Delegates responsibility for developing and implementing objectives and methods for design and construction programs. Applies VE to the activities within the main office and those of its regional offices and contractors. Keeps abreast of engineering requirements and status of construction programs by maintaining close contact with the engineering staffs. This close contact will be used to generate interest and cooperation in a continuing reappraisal of selected projects using the VE methodology. Analyzes requirements, technical instructions, and engineering reports of the directorate to determine possible changes that could generate savings.

2. Selects and assigns priorities for the application of VE based on the magnitude of potential savings. Participates in or monitors VE studies. Coordinates the application of VE studies with the heads of the other divisions of the organization. Plans, leads, and participates in conferences on recommended construction changes evolved through the VE program. Ability to gain approval of proposals from both design and engineering will have a marked influence on the success of attaining program objectives.

Maintains a close working relationship with counterparts in other agencies for the exchange of VE developments. If required, will attend professional meetings.

3. Promotes the use of VE throughout the program by encouraging contractors to participate. Conducts a continuing appraisal of VE progress in construction projects, by analysis of feedback information from the contractor to design, and by assisting in

the processing of VECPs.

4. Develops procedures for reporting VE information by the organization and field activities; correlates such information and disseminates it to various management levels. Publishes newsletters, highlighting VE activities.

Visits field organizations and other locations where the VE program is implemented or may be applied. Observes current methods and practices to ensure that they are conducive to efficient and economical prosecution of the program. Gives advice and formulates conclusions regarding the activities visited.

Lines of Responsibility

The value engineer works under the general supervision of the_____ who outlines broad objectives and general policy guidelines. Within these broad objectives and guidelines, decisions made by the incumbent are considered authoritative. Work is reviewed for overall effectiveness of final results.

Supervision of Others

Manages and evaluates engineers, technicians, and other personnel in their conduct of the VE program.

Specialized Qualification Requirements — Other Factors

The incumbent should possess a high degree of knowledge in fields of design and construction. To perform his duties satisfactorily, the incumbent must have enough initiative, self-organization, and perseverance to start and complete projects with little, if any, supervision or precedents. This position requires an engineering degree at the bachelor level.

Introduction

The following is a typical work outline for the presentation of a VE workshop seminar specifically oriented to architect-engineer services and the construction area.

WORK OUTLINE BREAKDOWN

Familiarization and Workshop Preparation

It is proposed to expend   * man-days over a maximum period of 10 days to:

1. Review and familiarize our VE staff with your organization and with selected project plans and specifications.
2. Finalize workshop agenda and assign your personnel to teams. (suggest 4 or 5 teams of 4 or 5 members, a maximum of 10 teams recommended)
3. Work jointly with designated personnel in selecting specific workshop projects and gathering support data.

During the preparation phase the workshop agenda will be completed and submitted, together with a set of training materials, to each attendee (maximum 25 sets). The training material is to be retained by the participants. Attached is a proposed workshop agenda prepared for your firm. Also, an itemized list of the required training aids and the time schedule for their use will be submitted.

Workshop

It is proposed to utilize two VE professionals fulltime during the course of study.† These professionals shall be experienced both in the instruction of VE methodology and in the actual preparation of VECPs applicable to the construction industry.

As outlined in the agenda, it is proposed to orient the first session of the workshop as a general briefing for workshop attendees and selected management personnel.

Instructors will work directly with the team to provide motivation, develop skills, and assist in gathering required information for

*Varies from 3 to 10 days depending on project size and complexity.

†If desired by the client, additional trained personnel can be supplied to supplement the staff as deemed necessary by project size and complexity.

meaningful proposal development.

The contractor's personnel will aim to motivate teams to generate potential savings in excess of seminar costs. Life-cycle costing will also be utilized for the development of potential savings.

The VE job plan to be used by workshop instructors is outlined as follows:

1. Information phase

   Define and evaluate function
   Establish worth

2. Speculative phase

   Develop alternates

3. Analytical phase

   Develop cost estimates
   Select alternates
   Evaluate alternates

4. Proposal phase

   Gather supporting data
   Prepare proposals
   Present proposals

The VE staff will be responsible for ensuring that teams develop all project worksheets and finished proposals in a format suitable for retention by the client.

Postcourse Evaluation

In the subsequent period, not to exceed 14 days, it is proposed that an additional   * man-days of effort be spent in a postcourse evaluation. During this time the course instructors will review and refine proposals.

Initially, workshop proposals will be reviewed with cognizant client personnel to gain their input. Subsequently, the contractor will assimilate these comments and refine the proposals. Required manufacturers' quotes, detailed cost data, and technical information that could not be collected during the time period of the workshop will be gathered for additional proposal documentation.

At this time, all follow-up data necessary to explain proposals to the reviewing and approving authorities will be collected.

*Varies from 3 to 10 days depending on project size and complexity.

Based on project observations and evaluations of client personnel comments (see Agenda, fifth session), the contractor, who has conducted similar evaluations for other organizations, will prepare a narrative report containing:

1. Recommendations for improvement of in-house training
2. General comments on the class participation and proposals developed
3. A list of potential VE projects isolated during contract for future investigation
4. Recommendations pertinent to a future VE program

VE WORKSHOP SEMINAR AGENDA

First session	**Objective: To introduce the value concept**
0800 - 0900	Registration
0900 - 0905	Introductory remarks

       Welcome (client speaker)
       Purpose of this course

0905 - 1200   Introduction to VE - For management and participants

       History and advantages of VE
       VE incentive contracting in defense department

          Purpose
          Advantages
          Methods
          Examples

       The organized approach

       Outline of the VE job plan
          Information phase
          Speculative phase
          Analytical phase
          Proposal phase

1200 - 1300   Lunch

1300 - 1400	Information phase: The functional approach

Identifying functions
Classifying functions
Setting value targets
Establishing cost/value ratios

1400 - 1600	Techniques of project selection

Areas of high cost
Evaluation criteria

1600 - 1700	Workshop: Information phase

Get acquainted with projects
Develop project information requirements
Define and classify project functions

Second Session	**Objective: Project workshop - information and speculative phase**

0800 - 0900	Film: Evaluation of function, cost, and worth
0900 - 1000	Techniques of cost analysis
1000 - 1200	Workshop: Apply cost analysis techniques to project
1200 - 1300	Lunch
1300 - 1700	Speculative phase: Developing alternates

Brainstorming
Blast-create-refine
New combinations

Invention and true creativeness
Nonjudicial approach
Sources of ideas

Third Session	**Objective: To complete the speculative phase and introduce the analytical phase**

0800 - 0900	Workshop: Develop list of alternate solutions

0900 - 1000   Analytical phase

       Evaluate basic function
       Evaluate by comparison
       Put a dollar sign on each idea
       Refine ideas

1000 - 1015   Break

1015 - 1200   Workshop: Analytical phase

       Analyze ideas
       Select best alternative ideas
       Investigation and evaluation

1200 - 1300   Lunch

1300 - 1700   Analytical phase (cont.)

       Investigation
        Introduction
        Consulting specialists
        Use new products and materials
        Challenge specifications and criteria

Fourth Session **Objective: To complete the analytical and
           introduce proposal phase**

0800 - 0945   Analytical phase (cont.)

       Evaluate ideas
       Rank ideas
       Select best alternate proposal

0945 - 1000   Break

1000 - 1100   Case studies

       Wall design
       Communications system example
       Other examples

1100 - 1200   Proposal phase

       Introduction
       Techniques

1200 - 1300	Lunch
1300 - 1700	Workshop: Proposal phase
	Review project work

Fifth Session	**Objective: Prepare and present proposals**
0800 - 1015	Workshop: Prepare proposal
	Develop sketches Develop implementation recommendations Develop report Prepare submittal format
1015 - 1030	Break
1030 - 1100	Review of presentations
	Team critiques
1100 - 1200	Project cleanup
1200 - 1300	Lunch
1300 - 1530	Presentation of proposals
1530 - 1700	Critique and closing remarks
	Attendees' comments on workshop Attendees' suggestions for future

APPENDIX H. GENERAL SERVICES ADMINISTRATION, PUBLIC BUILDINGS SERVICE, VALUE PROGRAMS FOR ARCHITECTS-ENGINEERS AND CONSTRUCTION MANAGERS

Extensive VE Services
for A-E Contracts

A. DESIGN CONCEPT STAGE

103. The A-E shall examine all design criteria for each of the various disciplines of work; i.e., site, architectural, structural, mechanical, and electrical, furnished under the contract, for the purpose of identifying and questioning constraints to achieving the required task at the lowest overall cost consistent with desired performance requirements.

   a. Based upon this examination, the A-E shall submit a report to the Contracting Officer identifying areas where criteria changes are considered desirable to develop maximum savings in structures, equipment, materials, or methods, even though such recommendations may be at variance with existing GSA criteria or instructions provided concerning the design in question.

   b. Recommended criteria modifications, together with the magnitude of savings therefor, should be included in the report. Use of GSA Form 2762, Design Review Ideas, is suggested.

   c. This report shall be submitted with the design concepts.

B. DESIGN TENTATIVE STAGE

111. The A-E shall perform a task team effort to review his tentative design submittal for VE ideas and cost effectiveness. Immediately upon completion of the task team effort, the A-E shall submit a report of the teams' VE ideas to the Contracting Officer. Use of GSA Form 2762, Design Review Ideas, is suggested.

112. The A-E shall host a 40-hour VE Workshop using a qualified VE Consultant. The Workshop shall be held after completion of the design tentative submittal.

   a. The Workshop shall be attended by a suitable number of employees from the A-E's firm, and from each design consultant firm. In addition, the A-E shall invite the attendance of representatives from the Government, the various using agencies of the facility under design and other individuals that may be recommended by the A-E and approved by the Government in order to achieve an optimum attendance of 25-30 people.

   b. The A-E, in concert with the Government, shall identify areas in the design for value study during the Workshop. The A-E shall complete the information phase of the VE Job Plan referenced in the VE Handbook and shall prepare a data package for each VE study to be used in the Workshop.

   c. The Workshop schedule, project studies, and program instruction shall be submitted to the Contracting Officer for approval at least 3 weeks in advance of the proposed commencement of the Workshop.

   d. The closing portion of the Workshop involving team project presentations shall be held in a suitable location to facilitate attendance by Government management officials. The A-E shall provide the Government with a copy of each teams' VE Workbook and Executive Brief developed during the Workshop.

e. The A-E shall provide suitable facilities, programs, and loose-leaf notebook binders for participants.

f. The Government will provide guest speakers as desired, VE Handbooks as reference material, and workshop completion certificates.

C. INTERMEDIATE WORKING DRAWING STAGE

120. The A-E shall continue to utilize the individuals of his firm and those of his design consultants who have received VE experience under this contract as a special value review team to provide comments, ideas, and recommendations for value improvement through the rest of the design effort.

121. The A-E shall perform a task team effort to review his intermediate working drawing design for VE ideas and cost effectiveness. This review should concentrate on high volume use items provided in the design such as doors, valves, finishes, etc. where the cost to make a change is minimal compared to the potential savings.

D. POST CONSTRUCTION CONTRACT AWARD STAGE

130. The Government will receive and initially screen VECPs submitted by the construction contractor(s) pursuant to the VE Incentive Clause in his contract.

a. The A-E will be asked to review and comment on those VECPs the Government is considering favorably.

b. If the A-E considers that adoption of a specific VECP requires design effort beyond the scope of this contract, the A-E shall so notify the Contracting Officer in accordance with Clause 3 of the General Provisions of this contract. The Contracting Officer will then determine whether or not the additional service will be required.

F. STANDARD SERVICES

141. The conducting of VE Workshops as required by this contract shall be performed by an outside VE Consultant. The A-E shall submit for approval of the Contracting Officer the name(s) of the VE Consultant (with their record of education and experience) whom he proposes to provide the workshop leadership required by this contract. The VE Consultant shall be qualified by education, training, and experience equivalent to that required by the Society of American Value Engineers for certified value specialists and shall have a recognized background in the field of Value Engineering.

142. The VE task team required by this contract shall be composed of at least one architect, structural engineer, mechanical engineer, and electrical engineer with other support as necessary. Individuals on this team should not be the same persons responsible for the original design work. The A-E shall submit for approval of the Contracting Officer the names of the members of the VE task team (with their record of education and experience) whom he proposes to perform the VE design reviews required by this contract. The A-E can meet this requirement by one of the following three options:

a. Utilize employees of the A-E and his design consultants, all of whom have completed a 40-hour VE Workshop course accredited by the Society of American Value Engineers or its equivalent.

b. Engaging a qualified VE Consultant to work with and guide the inexperienced employees of the A-E's firm and his design consultants. In this option the VE Consultant would serve as team leader.

c. Engaging a qualified VE Consultant to provide the full service. The VE Consultant shall be qualified by education, training, and experience equivalent to that required by the Society of American Value Engineers for certified value specialists and shall have a recognized background in the field of Value Engineering.

144. The A-E shall review and consider for incorporation, VE changes and ideas suggested by the Government, by his VE review team or by his VE Consultant. The A-E shall take one of the following courses of action for each idea suggested:

a. During design, incorporate the idea provided that it is convenient to do so at no additional fee and so advise the Government.

b. Reject the idea and advise the Government as to the reason.

c. Recommend approval of a change to completed and approved design work or GSA criteria and advise the Contracting Officer of the additional fee necessary. to incorporate the idea in accordance with Clause 3 of the General Provisions (SF 253) of this contract. The Contracting Officer will then determine whether or not the additional service will be required.

d. Use of GSA Form 2777, Summary Report, is suggested to report disposition of VE ideas.

147. The Government will furnish VE Workbooks, Executive Briefs, and all other referenced forms for the information and use of the A-E as necessary.

G. HANDBOOKS

150. VE Handbook, PBS P 8000.1, will be issued for the information and use of the A-E and his design consultants.

H. FEE AND PAYMENT

160. With regard to VE services, no incentive payments to or sharing of savings with the A-E will be made by the Government in connection with this contract.

VE Services for
Construction Manager Contracts

A. DESIGN PHASE

201. The CM shall examine all criteria for each of the various disciplines of work; i.e., site, architectural, structural, mechanical, and electrical for the purpose of identifying and questioning constraints to achieving the required task at the lowest overall cost consistent with desired performance requirements.

   a. Based upon this examination, the CM shall submit a report to the Contracting Officer identifying areas where criteria changes are considered desirable to develop maximum savings in structures, equipment, materials, or methods, even though such recommendations may be at variance with existing GSA criteria or instructions provided concerning the design in question.

   b. Recommended criteria modifications, together with the magnitude of savings, therefor, should be included in the report. Use of GSA Form 2692, Design Review Ideas, is suggested.

   c. This report shall be submitted with the design concept review.

202. The CM shall host a 40-hour VE Workshop using a qualified VE Consultant. The Workshop shall be held after completion of the A-E's design tentative submittal.

   a. The Workshop shall be attended by a suitable number of employees of the CM's firm. In addition, it will be attended by employees of the A-E's firm and his design consultants. The CM shall also invite the attendance of representatives from the Government and the various using agencies of the facility under design to provide for an optimum attendance of 25-30 people.

   b. The CM, in concert with the Government and the A-E, shall identify areas in the design for value study during the Workshop. The CM shall complete the information phase of the VE Job Plan referenced in the VE Handbook and shall prepare a data package for each VE study to be used in the Workshop.

   c. The Workshop schedule, project studies and program of instruction shall be submitted to the Contracting Officer for approval at least 3 weeks in advance of the proposed commencement of the Workshop.

   d. The closing portion of the Workshop involving team project presentations shall be held in a suitable location to facilitate attendance by Government management officials. The A-E shall provide the Government with a copy of each teams' VE Workbook and Executive Brief developed during the Workshop.

   e. The CM shall provide suitable facilities, programs, and loose-leaf notebook binders for participants in the Workshop session.

203. The CM shall continue to utilize the individuals of his firm who have received VE experience under this contract as a special value review team to provide comments, ideas and recommendations for value improvement through the rest of the design effort.

204. The CM shall perform a task team effort to review the intermediate working drawing design for VE ideas and cost effectiveness. This review should concentrate on high volume use items provided in the design such as doors, valves, finishes, etc. where the cost to make a change is minimal compared to the potential savings. Use of GSA Form 2762, Design Review Ideas, is suggested.

B. CONSTRUCTION PHASE

210. The CM shall encourage the Separate Construction Contractors on the project to participate in the VE program by submitting Value Engineering Change Proposals (VECPs) in accordance with the VE Incentive Clause in their construction contracts.

   a. As field representative of the Government, the CM will initially receive all VECPs submitted.

   b. The CM shall promptly review all VECPs and submit his recommendations regarding acceptance to the Contracting Officer. The CM shall suggest modifications to VECPs, when appropriate, in order to make the idea acceptable. The CM shall provide constructive, technical reasons for disapproval when he makes such a recommendation.

C. STANDARD SERVICES

220. The conducting of VE Workshops as required by this contract shall be performed by an outside VE Consultant. The A-E shall submit for approval of the Contracting Officer the name(s) of the VE Consultant (with their record of education and experience) whom he proposes to provide the workshop leadership required by this contract. The VE Consultant shall be qualified by education, training, and experience equivalent to that required by the Society of American Value Engineers for certified value specialists and shall have a recognized background in the field of Value Engineering.

221. The VE task team required by this contract shall be composed of at least one architect, structural engineer, mechanical engineer, and electrical engineer with other support as necessary. The CM shall submit for approval of the Contracting Officer the names of the members of the VE task team (with their record of education and experience) whom he proposes to perform the VE design reviews required by this contract. The CM can meet this requirement by one of the following three options:

   a. Utilize employees of the CM and his design consultants, all of whom have completed a 40-hour VE Workshop course accredited by the Society of American Value Engineers or its equivalent.

   b. Engaging a qualified VE Consultant to work with and guide the inexperienced employees of the CM Firm and his design consultants. In this option the VE Consultant would serve as team leader.

   c. Engaging a qualified VE Consultant to provide the full service. The VE Consultant shall be qualified by education, training, and experience equivalent to that required by the Society of American Value Engineers for certified value specialists and shall have a recognized background in the field of Value Engineering.

222. The Government will furnish VE Handbooks, Executive Briefs and all other referenced forms for the information and use of the CM as necessary.

223. The Government will provide guest speakers for Workshop sessions as desired, provide VE Handbooks as reference material, and will provide Workshop certificates for each participant.

G. HANDBOOKS

230. VE Handbook, PBS P 8000.1, will be issued for the information and use of the CM.

H. FEE AND PAYMENT

240. With regard to VE services, no incentive payments to or sharing of savings with the CM will be made by the Government in connection with this contract.

TABLE 1. HYPOTHETICAL OFFICE BUILDING DEVELOPMENT: ECONOMIC SUMMARY

Total construction cost	$26,933,000
Indirect costs	7,099,000
Total cost of improvements	34,032,000
Land cost	5,600,000
Total project cost	39,632,000
Less mortgage loan*	33,000,000
Equity investment	$ 6,632,000
Net operating income (10.3% of cost)	$ 4,080,000
Less debt service	3,046,000
Before tax-stabilized cash flow (15.6% of equity investment)	$ 1,034,000

---

*Loan amount
Net operating income $4,080,000
Capitalized @ 9.25% $44,108,000
Loan percent   x 0.75
Loan amount $33,000,000

Source: T.L. Karsten, Los Angeles, Calif. Paper presented at Construction Management Seminar, January 1972.

TABLE 2. HYPOTHETICAL OFFICE BUILDING DEVELOPMENT: AREA ANALYSIS

	Gross area(sq ft)	BOMA full-floor rentable area(sq ft)	BOMA divided-floor net rentable area(sq ft)
Building			
Ground floor	22,000	17,000	17,000
32 upper floors*	704,000	654,720	577,280
Mechanical penthouse	11,000		
Total office building	737,000	671,720	594,280

Parking garage

1 Parking space per 500 sq ft gross area = 1,474 spaces

6 Levels above grade @ 80,000 sq ft = 480,000 sq ft (325 sq ft/space)

Site

Building site	32,000 sq ft
Garage site	80,000 sq ft
Total site	112,000 sq ft

*Typical floor:
Gross area   22,000 sq ft
Full-floor rentable area 20,460 sq ft (93%)
Divided-floor net rentable area 18,040 sq ft (82%)

Source: T.L. Karsten, Los Angeles, Calif. Paper presented at Construction Management Seminar, January 1972.

TABLE 3. HYPOTHETICAL OFFICE BUILDING DEVELOPMENT: COST PROJECTION

Building, including tenant finish	737,000 sq ft @ $31.00	$22,847,000
Walkways and plaza	10,000 sq ft @ 5.00	50,000
Garage	480,000 sq ft @ 6.25	3,000,000
Subtotal		25,897,000
Contingency @ 4%		1,036,000
Total construction		26,933,000

Indirect costs:		
Architecture and engineering @ 4.75% + 0.5%	$1,414,000	
Taxes during construction	375,000	
Insurance during construction	20,000	
Interim financing — $33,000,000 @ 9% over 2 years as used	2,970,000	
Permanent loan fees — 1 point to lender, 0.5 point to broker	495,000	
Legal costs	50,000	
Closing costs	13,000	
Title insurance	48,000	
Completion bond	135,000	
Leasing commissions	432,000	
Tenant space layouts	40,000	
Construction supervision	100,000	
Development fees and overhead	675,000	
Initial losses and leasing concessions	332,000	
Total indirect costs		7,099,000
Total cost of improvements		34,032,000
Land	112,000 sq ft @ $50.00	5,600,000
Total project cost		$39,632,000

Source: T.L. Karsten, Los Angeles, Calif. Paper presented at Construction Management Seminar, January 1972.

TABLE 4. HYPOTHETICAL OFFICE BUILDING DEVELOPMENT: STABILIZED INCOME PROJECTION

Offices
  Owner occupied (initially)

Ground floor	17,000 sq ft @ $15.00	$ 255,000
16 upper floors	288,640 sq ft @ 8.75	2,526,000
Subtotal		2,781,000
Competitive rental space	288,640 sq ft @ 8.75	2,526,000
Gross potential income — Offices		5,307,000
Less vacancy on competitive space @ 5%		126,000
Effective gross income — Offices		5,181,000

Parking

1,032 Monthly spaces @ $45 + 5% oversell	$585,000	
442 Transient spaces @ $1.35 x 2½ turns x 250 days	373,000	
Total parking income		958,000
Total effective gross income		6,139,000

Operating and fixed charges
Building

Operating costs	594,280 sq ft @ $1.90	1,229,000
Property taxes	594,280 sq ft @ 1.00	594,000

Parking garage

Operating costs	1,474 spaces @ $110	162,000
Property taxes	1,474 spaces @ 50	74,000
Total operating and fixed charges		2,059,000
Net operating income	(10.3% of cost)	$4,080,000

All areas are divided-floor basis

Source: T.L. Karsten, Los Angeles, Calif. Paper presented at Construction Management Seminar, January 1972.

TABLE 5. HYPOTHETICAL OFFICE BUILDING DEVELOPMENT: CALCULATION OF INITIAL LOSSES AND LEASING CONCESSIONS (IN DOLLARS X 1000)

Month of Operation	Office income	Parking income	Total cash inflows	Office, oper. & fixed charges	Parking, oper. & fixed charges	Debt service	Total cash out flows	(Cash losses)
1	301	56	357	125	20	234	379	(22)
2	309	57	366	126	20	234	380	(14)
3	317	59	376	128	20	234	382	(6)
4	324	60	384	129	20	234	383	(1)
5	332	62	—	131	20	234	—	—
6	340	63	—	133	20	234	—	—
7	347	65	—	135	20	234	—	—
8	355	66	—	136	20	234	—	—
9	363	67	—	138	20	234	—	—
10	370	68	—	140	20	234	—	—
11	378	69	—	141	20	234	—	—
12	386	71	—	143	20	234	—	—
13	393	72	—	145	20	234	—	—
14	401	74	—	146	20	234	—	—
15	409	76	—	147	20	234	—	—
16	416	77	—	149	20	234	—	—
17	424	79	—	150	20	234	—	—
18	432	80	512	152	20	234	406	—

Total initial losses   43

Add leasing concessions   est. @ $1.00 sq ft on competitive rental space of 288,640 sq ft.   289

Total initial losses and leasing concessions   S 332

Assumptions: Owner-occupied space and 33% of competitive rental space is leased upon opening; 18 months to reach stabilized occupancy. Parking income increases pro rata with building income. Operating and fixed charges on vacant office space are projected at $1.75/sq ft during fill-up; parking garage operating and fixed charges are assumed to begin at the stabilized level. Interest only payments to mortgage loan in first 18 months.

Source: T.L. Karsten, Los Angeles, Calif., Paper presented at Construction Management Seminar, January 1972.

APPENDIX J. VALUE ENGINEERING REQUIREMENTS FOR CONSTRUCTION MANAGEMENT CONTRACTS

Definition

Value Engineering is an organized study of project requirements and details by a multi-disciplined group of professionals, aimed solely at reducing unnecessary initial and ownership costs.

Scope of Work For A Typical Project

Value Engineering services consist of the following:

Conceptual

As early as feasible during the conceptual stage, a meeting is held between key project design personnel and the construction management (CM) staff. During the meeting key decisions effecting major cost items are reviewed. All Value Engineering information developed during previous projects of similar nature is provided to the designers. Subsequently, a Value Engineering review is conducted on the project criteria. The Value Engineering team identifies potential areas of initial and/or life cycle savings due to project criteria, owner requirements, etc. Based on the results, a report is submitted to the owner/designer for review and implementation of acceptable savings.

Preliminary Review

During the development of the project requirements (design concept), all available project documents are assembled, including owner requirements, design calculations, cost data, preliminary plans and outline specifications. A multi-disciplined team consisting of trained professionals in civil, architectural, structural, mechanical and electrical areas is gathered to review project documents. The team uses a job plan. Initially, an "idea listing" suggesting areas of potential savings or increased benefit is developed together with ranges of savings. The idea listing resulting from the "brainstorming" session is reviewed with representatives of the owner and designer. Pertinent remarks are recorded and follow-up actions delineated for implementing savings. Typical follow-up actions include in-depth VE studies on major areas as deemed applicable by the CM.

Ideally, the project review is conducted after architectural schematics are substantially completed and before the major mechanical and electrical decisions are fixed. The major objective of this phase is to optimize interdisciplinary cost impacts.

A final report documenting the above is forwarded to the owner.

Working Drawings Review

During the early stages of the architectural working drawings and just after the initial submissions of mechanical and electrical working drawings, a second review of the project should be conducted. All available project documents are assembled such as: plans, draft specifications including general conditions, cost data, and structural, mechanical, and electrical design calculations.

As in Preliminary Review above, an idea listing is developed. During this review the team will also isolate areas which will improve document interpretation, assure greater completeness, and result in greater cost effectiveness, but which may not necessarily reduce initial costs of each bid package. The Value Engineering team will focus primarily on a review of materials and methods, specification context, and other detailed contract requirements.

Review of the idea listing is conducted by the Value Engineering team and the owner and designer. In depth Value Engineering studies will be conducted on major areas having significant cost savings as deemed applicable by the CM. As in Preliminary Review, above a written report summarizing the results of the study is submitted to the owner.

Construction Contract

As a standard procedure the construction manager recommends inclusion of Value Engineering incentive contract provisions for all contracts over $100,000. The construction manager's Value Engineering team will be available for review for contractor generated Value Engineering Change Proposals as deemed necessary by the project manager. In addition, wherever applicable the Value Engineering team offers services, on an informal basis as requested by the project manager, to the contractor to review cost savings proposals before submittal.

BIBLIOGRAPHY

Building Research Advisory Board, Federal Construction Council, *Value Engineering in Federal Construction Agencies*, Symposium-Workshop Report Number 4. May 27, 1969.

> Printing and Publishing Office
> National Academy of Sciences
> 2101 Constitution Avenue
> Washington, D.C. 20418
> ($3.50 per copy)

DoD Handbook 5010.8-H (September 12, 1968).

> Superintendent of Documents
> 710 North Capitol Street, N.W.
> Washington, D.C. 20402
> ($1.00 per copy)

General Services Administration, *Value Engineering* (Handbook). PBS P 8000.1 (Jan. 12, 1972) and Change .1 (March 2, 1973)

> GSA Business Service Centers
> Available at twelve GSA Service Centers
> in major locations
> ($5.00 per copy)

Miles, L.D. *Techniques of Value Analysis and Engineering*, 2nd ed. New York, McGraw-Hill, 1961.

> McGraw-Hill Book Company
> Book Distribution Center
> Hightstown, New Jersey 08520
> ($12.75 per copy)

INDEX

ABOUT THE AUTHOR

A. J. Dell'Isola, P.E., Vice-President and Director of the Value Management Division of Smith, Hinchman. & Grylls Associates, Inc., is a graduate of the Massachusetts Institute of Technology. He has been a consultant in value analysis/engineering (VA/E) to the construction industry since 1966 and has conducted over 75 VE services contracts for various private and municipal organizations and agencies. His prior experience includes 15 years of supervising heavy construction in the United States and abroad. In 1962 he was assigned as a Special Assistant for VE for the Naval Facilities Engineering Command and subsequently for the Office of the Chief of Engineers in Washington, D.C. During that time he pioneered a a VE program resulting in over $10 million in audited annual savings.

He has presented VE briefings to both the United States Senate and the House of Representatives. In 1969 he was appointed VA/E consultant to the President's Advisory Council on Management Improvement.

He has conducted numerous seminars for various agencies, universities, and technical societies, and has written several articles, published in the United States and abroad. Recently, he conducted a VA/E Workshop in Tokyo. Presently, he is a guest lecturer for Advanced Management Research, New York, conducting both national and international lectures on VE in construction.

The *Engineering News-Record* cited him in 1964 for outstanding achievement in value engineering, and in 1968 he received the Distinguished Service Award from the Society of American Value Engineers.